QUANTUM COMPUTING

for

High School Students

by

YULY BILLIG

QUBIT PUBLISHING
2018

Published by Qubit Publishing
23 Confederation Private, Ottawa, Ontario, K1V 9W6, Canada
http://qubitpublishing.com

Quantum Computing for High School Students

Credits

Cover design and artwork: Maria Zaynullina

Layout and typesetting: LaTeX

ISBN 978-1-7753904-0-4 (Hardcover)
ISBN 978-1-7753904-1-1 (Paperback)

Contents

Preface

The purpose of this book is to make the subject of quantum computing accessible to anyone with a knowledge of high school algebra and
trigonometry.

The theory of quantum computing combines quantum mechanics,
abstract algebra, computer science and cryptography. We made this
book self-contained by introducing various topics from these diverse
areas, needed for understanding of quantum computing.

The field of quantum computing is still very young. Quantum
computers that we are able to build today are not powerful enough to
solve practical problems, and to-date we have developed only a handful
of quantum algorithms. Yet, quantum computing holds great promise.
We know that large scale quantum computers have computational
power unmatched by the kind of computers we use now. Even though
these large scale quantum computers are technologically unattainable
today, the theory behind these devices is solid.

In 1994, Peter Shor discovered a quantum algorithm which will
allow one to break cryptography used today in Internet communications, once large scale quantum computers become a reality. Rigorous
exposition of Shor's algorithm is the central goal of this book.

Proper description of quantum mechanics requires complex numbers and complex vector spaces. In order to make presentation of
the theory more accessible, we avoid using complex numbers in this
book. This simplification still allows us to convey all significant ideas
of quantum computing, while making it much easier to visualize quantum states and quantum gates. In the last chapter, we briefly touch
upon the aspects of the theory left outside the scope of this book.

I am grateful to my father, Vladimir Billig, for the idea of writing a
book on quantum computing for high school students. I thank Anand
Srinivasan for suggesting quantum computing as a topic for a math
enrichment course. The course "Quantum Computing" was offered
by the Math Enrichment Centre at Carleton University in 2017/18,
and this book is based on my lecture notes. I thank the high school
students who took this course for their enthusiasm and hard work.

1 The Splendors and Miseries of Quantum Computers

Moore's Law states that the number of transistors in computers we can build doubles every two years. This progress is only possible if we make transistors ever smaller. In 2017, the width of a transistor is at the scale of 10 nanometers, which corresponds to a layer of only 50 atoms in depth. Already at this scale, quantum effects, such as quantum tunnelling, become significant. Clearly with the trajectory of Moore's Law, our present paradigm for computer architecture will soon hit the wall, since the size of a transistor cannot possibly be smaller than the distance between atoms in a crystal. Moreover as we approach this barrier, quantum effects will become more prominent.

In our daily life, we deal with the objects that consist of many atoms (their number in a grain of sand is 10^{20}). In large collections of atoms, quantum effects get averaged out, and as a result we do not experience quantum mechanics with macroscopic objects. Yet quantum mechanics is increasingly present in our technology – such an ordinary thing like an LED flashlight, operates on quantum principles.

The idea of quantum computing is to embrace the bizarre quantum world, instead of fighting its influence. This is not easy, but there is a lot to gain. Quantum computers are devices that use quantum systems as processors.

What are the riches offered by quantum computers?

1. We get exponentially more memory, compared to our present computers.

2. We will be able to run massively parallel computations, again exponentially more parallel than anything we can envision with classical computers.

What are the challenges?

1. There is no direct access to memory. The act of reading from quantum memory has a probabilistic outcome and destroys the records as they are being read.

2. The quantum processor should be fully isolated from the environment, yet we should have access to it to control it.

3. We do not yet fully understand how to write efficient quantum algorithms which take advantage of the power of quantum computers.

Quantum computers have sound theoretical foundations in both physics and mathematics. However technological obstacles remain very serious, and a significant breakthrough is required. A lot of progress is also needed in developing quantum algorithms. In order to work on algorithms, one does not need access to a quantum computer, but only pen and paper, empowered with the knowledge of the theory of quantum computing.

Let us try to understand the difference between classical and quantum computers. In a classical computer data is stored in the memory as sequences of 0's and 1's. The unit of memory is called the bit, and it can store either 0 or 1. For the purpose of this discussion, it is useful to view 0 and 1 purely as symbols.

The unit of memory of a quantum computer is called the *qubit*, and it can store 0 and 1 simultaneously. More precisely, the value of a qubit is a vector with length 1 on a plane:

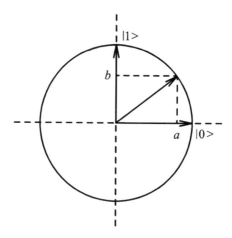

In order to make a connection with a classical bit, we label one coordinate axis with symbol 0 and the other axis with symbol 1. Accordingly, for the unit vectors on the coordinate axes we use notations which are traditional in quantum mechanics:

$$|0\rangle = \begin{pmatrix} 1 \\ 0 \end{pmatrix}, \quad |1\rangle = \begin{pmatrix} 0 \\ 1 \end{pmatrix}.$$

A qubit may be then written as a vector

$$\begin{pmatrix} a \\ b \end{pmatrix} = a\,|0\rangle + b\,|1\rangle,$$

which is interpreted as a superposition of two classical bit values 0 and 1 with the weights a and b, where $a^2 + b^2 = 1$. We emphasize that $|0\rangle$ is not a zero length vector, but rather a unit length vector on an axis that is labelled with symbol "0".

When we build a computer as a physical machine, we need to use physical objects which can implement our abstract constructions of a bit and a qubit. A capacitor (an electronic device that can hold an electric charge) may serve as a unit of memory of a classical computer. A charged state of a capacitor represents 1, while discharged state represents 0.

A photon may serve as a physical realization of a qubit. A photon is a quantum of an electromagnetic wave. Imagine a photon flying in a 3-dimensional space along the Z-axis. As it propagates, electric field and magnetic field oscillate in mutually perpendicular directions in XY-plane.

The specific way how this oscillation occurs, is called the *polarization* of a photon. There are two kinds of polarization – circular and linear. In circular polarization, the electric field spirals around the Z-axis as the photon propagates. In this book we will only consider a simpler case of a linear polarization, when the electric field oscillates in a fixed direction perpendicular to the Z-axis. Linearly polarized photons may be obtained by passing a beam of light through a polarizing filter.

Polarized light is used in 3D movies. To create a 3D effect, left and right eyes should see slightly different pictures. The movie is shot with two cameras that are slightly apart. The images from both cameras are simultaneously projected on the movie screen, but the light from

4

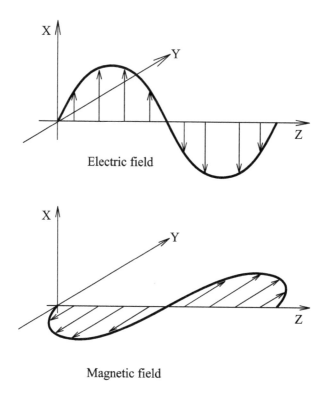

Electric field

Magnetic field

the two projectors are polarized in two different ways. The glasses
have polarizing filters, each passing light only from one projector. As
a result, two eyes receive distinct pictures, creating a 3D effect.

Imagine a photon with a linear polarization at an angle α in XY-
plane. This photon can be used as a physical implementation of a
qubit with value

$$\cos(\alpha)\,|0\rangle + \sin(\alpha)\,|1\rangle\,.$$

A small technical point about the photon states $|0\rangle$ and $-\,|0\rangle$.
Both states correspond to photons with the same axis of polariza-
tion, however the oscillation of the electric field for $-\,|0\rangle$ occurs in
antiphase relative to $|0\rangle$. Individually, these photons are essentially
indistinguishable, however given a pair of such photons, we can detect
the difference in phases, and view their states as two distinct qubit

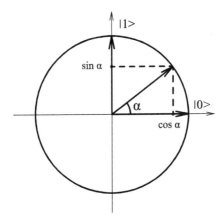

values.

Now let us progress towards multi-qubits. A 2-qubit is a vector with 4 components of the form:

$$a_0 \left|00\right\rangle + a_1 \left|01\right\rangle + a_2 \left|10\right\rangle + a_3 \left|11\right\rangle,$$

where $a_0^2 + a_1^2 + a_2^2 + a_3^2 = 1$. The basis vectors in the space of 2-qubits, $\left|00\right\rangle$, $\left|01\right\rangle$, $\left|10\right\rangle$, $\left|11\right\rangle$, also called *pure states*, correspond to the 4 possible classical values of 2-bit expressions. A general 2-qubit stores a combination (also called *superposition*) of the 4 classical 2-bit values simultaneously, with weights.

For example, the 2-qubit $0.3 \left|00\right\rangle + 0.1 \left|01\right\rangle + 0.9 \left|10\right\rangle + 0.3 \left|11\right\rangle$ is a superposition of all 4 classical values, but the 2-bit value "10" has a heavier weight in this 2-qubit.

As you might now guess, a 3-qubit is a vector with 8 components:

$$a_0 \left|000\right\rangle + a_1 \left|001\right\rangle + a_2 \left|010\right\rangle + a_3 \left|011\right\rangle$$
$$+ a_4 \left|100\right\rangle + a_5 \left|101\right\rangle + a_6 \left|110\right\rangle + a_7 \left|111\right\rangle.$$

Notice the pattern in our notations between the index of the coefficient "a" and the label of the corresponding pure state. Take the term $a_6 \left|110\right\rangle$, for example. Here "110" is the binary expression for the integer 6.

We can see that the number of terms in these expressions doubles with each additional qubit. Thus for an n-qubit the number of terms will be 2^n. An 8-qubit involves 256 terms:

$$a_0\left|00000000\right\rangle + a_1\left|00000001\right\rangle + a_2\left|00000010\right\rangle + \ldots + a_{255}\left|11111111\right\rangle.$$

As an exercise, let us determine the index of "a" for the term with $\left|11010011\right\rangle$ in this expansion. The digits in a binary expansion correspond to powers of 2 (as opposed to powers of 10 in the decimal form). For an 8-bit expression, the leftmost digit corresponds to 2^7, while the rightmost digit corresponds to 2^0. We read the binary expression "11010011" as an integer

$$1 \times 2^7 + 1 \times 2^6 + 0 \times 2^5 + 1 \times 2^4 + 0 \times 2^3 + 0 \times 2^2 + 1 \times 2^1 + 1 \times 2^0$$
$$= 128 + 64 + 16 + 2 + 1 = 211 \text{ (decimal)}.$$

Conversely, in order to write 117 (decimal) in a binary 8-bit form, we expand 117 as a sum of powers of 2:

$$117 = 64 + 53 = 64 + 32 + 21 = 64 + 32 + 16 + 5 = 64 + 32 + 16 + 4 + 1$$
$$= 0 \times 2^7 + 1 \times 2^6 + 1 \times 2^5 + 1 \times 2^4 + 0 \times 2^3 + 1 \times 2^2 + 0 \times 2^1 + 1 \times 2^0$$
$$= 01110101 \text{ (binary)}.$$

In order to use more compact notations, we shall sometimes write an 8-qubit using a decimal form:

$$a_0\left|0\right\rangle + a_1\left|1\right\rangle + a_2\left|2\right\rangle + \ldots + a_{255}\left|255\right\rangle.$$

Here we understand that all decimal integers appearing in the notations of basis vectors need to be converted to the 8-bit binary form.

As the number of bits increases, such expressions will become very long. An efficient mathematical way of writing such sums is to use the Σ notation. With this notation an 8-qubit is written compactly as

$$\sum_{k=0}^{255} a_k\left|k\right\rangle.$$

Here k is the index of summation and runs from 0 to 255, so the sum has 256 terms. When $k = 0$, it produces the summand $a_0 |0\rangle$, $k = 1$ yields $a_1 |1\rangle$, and so on. For each basis vector $|k\rangle$, the integer k is understood to be in an 8-bit binary form.

As we shall see in the next chapter, the joint polarization state of n interacting photons is described as an n-qubit. The amount of memory required to record such a state on a classical computer grows exponentially in n. If we allocate 1 byte to record the value of each "a" coefficient, then we need 2 bytes to store a 1-qubit, one kilobyte to store a 10-qubit, one megabyte to store a 20-qubit, one gigabyte to store a 30-qubit, one terabyte to store a 40-qubit, one petabyte to store a 50-qubit. If we take all the matter in the visible Universe, and make a giant memory chip based on today's approach to computer memory, we will not be able to store a 100-qubit on that device. At the same time, a collection of 100 interacting photons is something that may be everywhere around us. This realization led to inception of quantum computing.

The idea of using quantum systems as computational devices was put forward in 1980 independently by Yuri Manin and Paul Benioff. This idea was also discussed by Richard Feynmann in 1982. Foundations of this theory were systematically developed by David Deutsch, but a real explosion in this area was caused by the discovery of Shor's algorithm in 1994. The quantum algorithm that Peter Shor has developed, will break most of the public key cryptography which we use today in Internet communications, once large-scale quantum computers are built. The goal of this book is to explain Shor's algorithm and all the background material required for understanding it.

8

2 Quantum Mechanics Demystified (not really)

In this chapter we will discuss axiomatics of quantum mechanics. Since Newton, motion of physical objects has been described with numerical quantities, such as position, velocity, acceleration, and physical theories gave a precise prediction of the future trajectory of an object if all the forces as well as the initial conditions are known. Quantum mechanics makes a departure from this certainty. It postulates that we can only predict probabilities for the future events, and that uncertainty is inherent in the laws of nature.

The *state* of a quantum system is a vector of length 1 in the *space of states*. For the purposes of quantum computing, we will take this space to be the space of n-qubits, for a particular value of n. The space of n-qubits is 2^n-dimensional since each n-qubit has 2^n components, just like a 3-dimensional vector has 3 components. In general, in quantum mechanics the space of states can be of any dimension, and even infinite-dimensional. Moreover, in the proper formulation of quantum mechanics, the components of vectors should be taken to be complex numbers, however we will restrict all components to be real, for the sake of simplicity.

In classical mechanics, a trajectory of an object is described with the Newton's 2nd Law, which is mathematically written as a differential equation. Newton's 2nd Law expresses the value of the acceleration of the object (mass × acceleration = force), and acceleration is the second derivative of the position, hence we get an equation on the second derivative, which makes it a differential equation. In quantum mechanics, evolution of a quantum state is also governed by a certain differential equation, called the Schrödinger equation. The Schrödinger equation itself is not essential for us, and we will not write it down here, but what is important is the fact that evolution of closed quantum systems is given by *orthogonal linear transformations*. We will defer explaining what exactly these transformations are, to a later chapter.

Probabilistic nature of quantum mechanics is exhibited in the process of *measurement*. Measurement is the only way to extract data from a quantum state. Given an n-qubit

$$\sum_{k=0}^{2^n-1} a_k \left| k \right\rangle$$

with the normalization (length 1 condition)

$$\sum_{k=0}^{2^n-1} a_k^2 = 1,$$

when we perform a measurement on it, we will obtain one of the classical n-bit values from $000\ldots0$ to $111\ldots1$, where the value k (in binary form) will appear with probability a_k^2. The normalization condition says that the sum of probabilities is 1, as it should be.

In order to make a measurement, we have to actively interfere with our quantum system. As a result, once the measurement is performed, the quantum state is destroyed. After the measurement the quantum system goes into the pure state corresponding to the observed value.

For example, if we perform a measurement on the 2-qubit

$$0.3 \left| 00 \right\rangle + 0.1 \left| 01 \right\rangle + 0.9 \left| 10 \right\rangle + 0.3 \left| 11 \right\rangle,$$

there may be 4 different outcomes:

- With probability 9% we observe "00" and the 2-qubit goes into the pure state $\left| 00 \right\rangle$.

- With probability 1% we observe "01" and the 2-qubit goes into the pure state $\left| 01 \right\rangle$.

- With probability 81% we observe "10" and the 2-qubit goes into the pure state $\left| 10 \right\rangle$.

- With probability 9% we observe "11" and the 2-qubit goes into the pure state $\left| 11 \right\rangle$.

Sometimes we may want to measure only some of the qubits in an n-qubit. Let us discuss what happens in this case. Suppose we have a 3-qubit

$$0.3\,|000\rangle - 0.6\,|001\rangle - 0.1\,|010\rangle - 0.7\,|011\rangle + 0.1\,|101\rangle - 0.2\,|110\rangle$$

and we measure the value of its first two bits, but not the third. There are four possible values of the first two bits that we can observe: 00, 01, 10 and 11. What are the probabilities of observing each outcome? To find it, we add the squares of coefficients of all terms with the given values of the first bits. After the measurement the first two bits will assume definite values, the ones that have been observed. The value of the third bit will not be fixed, however. To determine new state, we keep only the terms of the qubit that correspond to the observed values, and then renormalize the result, so that the new vector has length 1. For the above example we will have:

- With probability $0.09 + 0.36 = 45\%$ we observe "00" and the new state is $1/\sqrt{0.45}(0.3\,|000\rangle - 0.6\,|001\rangle)$.

- With probability $0.01 + 0.49 = 50\%$ we observe "01" and the new state is $1/\sqrt{0.5}(-0.1\,|010\rangle - 0.7\,|011\rangle)$.

- With probability 1% we observe "10" and the new state is $|101\rangle$.

- With probability 4% we observe "11" and the new state is $-\,|110\rangle$.

Let us describe how a measurement procedure may be implemented with the polarizing filters. We can use a mirror-like polarizing filter which reflects the light with the horizontal polarization $|0\rangle$, and lets through the light with the vertical polarization $|1\rangle$. What happens when a photon that is linearly polarized at an angle α is sent through this filter? We cannot predict what will happen to this photon, the outcome of this experiment is probabilistic. Since the initial state of this photon is

$$\cos(\alpha)\,|0\rangle + \sin(\alpha)\,|1\rangle,$$

with probability $\sin^2(\alpha)$ it will pass through the filter and will come out with the vertical polarization $|1\rangle$, and with probability $\cos^2(\alpha)$ it will get reflected and will change its polarization to horizontal, $|0\rangle$.

If we take out the filters from the 3D movie glasses, and put the two polarizing filters against each other, then we will notice that in one alignment they will be fairly transparent, but when one of the filters is rotated $90°$ relative the other, virtually all light will be blocked. As we rotate the filter, the transparency gradually changes, and the prediction from the discussion above is that the intensity of the light that goes through is proportional to $\cos^2(\alpha)$, where α is the angle between the axes of the filters.

The general scheme of a quantum computation consists of 3 steps:

1. Initialization. Quantum computer is initialized to a value of n-qubit that encodes the input of the quantum algorithm, or simply to $|00\ldots0\rangle$ if the input is built into the algorithm itself.

2. Running the Quantum Algorithm. This is essentially a complicated orthogonal linear transformation. Just as classical computations are broken down into elementary operations with binary logic, quantum algorithms are also split into elementary quantum operations, each involving only one or two quantum bits.

3. Performing the measurement of the final quantum state.

Here we can see the main difficulty in designing quantum algorithms. The final step of the computation is probabilistic, and in principle may produce any output. The quantum algorithm needs to create a state, which after the measurement will produce a correct answer to the problem being solved with a high enough probability. Since the quantum state is changed after the measurement into a pure state which corresponds to the observed value, we have only one attempt at accessing the information recorded in the quantum state. If the answer we get is incorrect (fortunately in many important problems there is a simple way of checking with a classical computer whether a given answer is correct), then our only recourse is to re-run the quantum algorithm.

Next we are going to discuss *entanglement*, which is an important phenomenon in quantum mechanics.

Suppose that we have two qubits $a_0 |0\rangle + a_1 |1\rangle$ and $b_0 |0\rangle + b_1 |1\rangle$. We would like to join them together and form a 2-qubit with them.

This can be done using the *tensor product* operation:

$$(a_0\,|0\rangle + a_1\,|1\rangle)(b_0\,|0\rangle + b_1\,|1\rangle) = a_0 b_0\,|00\rangle + a_0 b_1\,|01\rangle + a_1 b_0\,|10\rangle + a_1 b_1\,|11\rangle.$$

Here we just expanded the left hand side and used concatenation to multiply basis vectors, e.g. $|0\rangle\,|1\rangle = |01\rangle$.

Physically this corresponds to taking two non-interacting photons and considering them as parts of a single quantum system.

Can we do the opposite and factor a 2-qubit as a product of two 1-qubits? Let us try to do this for the 2-qubit $\frac{1}{\sqrt{2}}\,|00\rangle + \frac{1}{\sqrt{2}}\,|11\rangle$:

$$\frac{1}{\sqrt{2}}\,|00\rangle + \frac{1}{\sqrt{2}}\,|11\rangle = (a_0\,|0\rangle + a_1\,|1\rangle)(b_0\,|0\rangle + b_1\,|1\rangle).$$

Equating the coefficients of the basis vectors, we get 4 equations:

$$a_0 b_0 = \frac{1}{\sqrt{2}}, \; a_1 b_1 = \frac{1}{\sqrt{2}}, \; a_1 b_0 = 0, \; a_0 b_1 = 0.$$

Multiplying the fist two equations together we get $a_0 a_1 b_0 b_1 = \frac{1}{2}$, while multiplying the last two equations, we get $a_0 a_1 b_0 b_1 = 0$, which is a contradiction. Hence, the 2-qubit $\frac{1}{\sqrt{2}}\,|00\rangle + \frac{1}{\sqrt{2}}\,|11\rangle$ cannot be factored as a tensor product of two 1-qubits. Such quantum states are called *entangled*. Parts of an entangled quantum systems cannot be considered separately. From the physical point of view, this means that the photons in the pair $\frac{1}{\sqrt{2}}\,|00\rangle + \frac{1}{\sqrt{2}}\,|11\rangle$ are interacting.

Today we can generate pairs of entangled photons. We have less success, however, with entangling pre-existing photons "on-the-fly".

It is due to entanglement that quantum computers have an astronomical memory capacity, and efficient quantum algorithms must use entangled states.

Let us figure out the distinction between non-entangled and entangled states with respect to the measurement. Consider the following experiment: pairs of photons are generated with one photon in each pair sent to location A (Alice) and the second to location B (Bob).

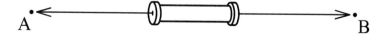

Alice and Bob perform measurements on the photons they receive.

Let us first start with the case when all pairs have the same unentangled state

$$a_0 b_0 \ket{00} + a_0 b_1 \ket{01} + a_1 b_0 \ket{10} + a_1 b_1 \ket{11}$$

as computed above. Let us determine the probability that Alice observes value "0" as a result of her measurement. Alice's value "0" corresponds to two pure states: $\ket{00}$ and $\ket{01}$, and hence this probability equals $a_0^2 b_0^2 + a_0^2 b_1^2 = a_0^2$.

Next, let us separate all observations into two groups – those where Bob observed "0", and those where he observed "1". Let us determine the probability for Alice to observe "0" within the first group. Such probability is called *conditional* probability, since we are looking for the probability of Alice's outcomes of observations which are conditional on Bob's outcome being "0".

Overall, the probability for both Alice and Bob to observe "0" is $a_0^2 b_0^2$, and the probability for Alice to observe "1", while Bob observes "0" is $a_1^2 b_0^2$. Thus within the group of observations where Bob observes "0", the ratio of Alice's "0"s to "1"s is $a_0^2 b_0^2 : a_1^2 b_0^2 = a_0^2 : a_1^2$. Since $a_0^2 + a_1^2 = 1$, this means that the conditional probability for Alice to observe "0" under the condition that Bob observes "0" is still a_0^2. Thus conditional probability in this case is equal to unconditional probability.

Two events, X and Y, are called *independent*, if the probability of X happening is equal to the conditional probability of X happening, under condition that Y happens as well.

Next consider the same experiment as above, where all pairs of photons are in an entangled state $\frac{1}{\sqrt{2}} \ket{00} + \frac{1}{\sqrt{2}} \ket{11}$. Here with probability 50% both Alice and Bob will register "0" and with 50% probability they will both register "1". If Bob observes "0" then the quantum state will collapse to $\ket{00}$ and Alice's result of measurement is guaranteed to be "0". While unconditional probability for Alice to observe "0" is 50%, the conditional probability of observing "0" under condition that Bob also observes "0" is 100%. What we see here is that for the entangled states, the outcomes of measurements for Alice and Bob are *correlated*, while for unentangled states the outcomes are independent.

In 1935 Einstein, Podolsky and Rosen wrote a paper about the foundations of quantum mechanics, where they expressed opinion that for entangled pairs of particles that are spatially separated, the result of measurement performed on the first particle cannot affect the state of the second particle. This would mean that outcomes of simultaneous measurements on spatially separated particles should be independent. Later experiments showed that they were wrong.

3 Cryptography: from the Roman Empire to Quantum Methods

In the 1970s, the invention of *public key cryptosystems* caused revolution in cryptography. Nowadays public key cryptography enables Internet commerce. One of our goals is to explain how quantum computing may be used to break public key cryptosystems; we will discuss this towards the end of the book. In this chapter, we will review older methods – secret key cryptography, and we shall see how the use of quantum mechanics could make them more efficient.

Julius Caesar was the first to employ cryptography systematically. He encrypted his military communications using what is now known as Caesar's cypher. When writing his messages, he would shift each letter by 3 positions in the alphabet backwards. Letters at the beginning of the alphabet were pushed in a cycle to the end of the alphabet:

A B C D E F G H I J K L M N O P Q R S T U V W X Y Z

X Y Z A B C D E F G H I J K L M N O P Q R S T U V W

so the text JULIUS CAESAR will be encoded as GRIFRP ZXBPXO.

We can give a mathematical interpretation of this cypher using the concept of remainders. Consider all possible remainders of integers after division by some fixed integer m (for the applications to cryptography in this chapter, we set $m = 26$). This set is denoted as $\mathbb{Z}_m$ and contains the remainders between 0 and $m - 1$. It is not difficult to see that in order to compute the remainder of a sum of two integers, it is sufficient to know only the remainders of the summands, and not the summands themselves. The same applies to multiplication as well.

Let us prove this simple fact. First we note that two integers a and b have the same remainder after division by m if the difference $a - b$ is divisible by m. Suppose we add two pairs of integers $a_1 + a_2$ and $b_1 + b_2$. We need to show that if a_1 and b_1 have the same remainder and a_2 and b_2 have the same remainder, then $a_1 + a_2$ and $b_1 + b_2$ also have the same remainder. Indeed, consider the difference:

$$(a_1 + a_2) - (b_1 + b_2) = (a_1 - b_1) + (a_2 - b_2).$$

Both summands in the right hand side are divisible by m, which implies our claim.

The argument for the product is similar:

$$a_1 a_2 - b_1 b_2 = (a_1 - b_1)a_2 + b_1(a_2 - b_2),$$

and we arrive at the same conclusion.

This allows us to define addition and multiplication of remainders in $\mathbb{Z}_m$: to compute the sum (resp. product) of two remainders, add (resp. multiply) them as integers, and take the remainder of the result. For computations with remainders, we put notation "mod m" at the end of the line, to indicate that these are the equalities of the remainders, and not the integers:

$$23 + 10 = 7 \mod 26,$$

$$23 \times 10 = 22 \mod 26.$$

Now to apply this to Caesar's cypher, we assign to each letter a remainder from $\mathbb{Z}_{26}$: A-1, B-2, ..., X-24, Y-25, Z-0. Then encryption is done by subtracting 3 from each letter code. Decryption is performed by adding 3 in $\mathbb{Z}_{26}$.

Caesar could use this simple method since this was a novel idea at the time, and his enemies were uneducated. Stronger Vigenère cypher came to replace Caesar's method. In Vigenère cypher, one chooses a secret word as a key for encryption and decryption. To illustrate this method, let us use the word QUBIT as a secret key to encrypt the message CHANGE PHOTON POLARIZATION:

C	Q	3	17	20	T
H	U	8	21	3	C
A	B	1	2	3	C
N	I	14	9	23	W
G	T	7	20	1	A
E	Q	5	17	22	V
P	U	16	21	11	K
H	B	8	2	10	J
O	I	15	9	24	X
T	T	20	20	14	N
O	Q	15	17	6	F
N	U	14	21	9	I
P	B	16	2	18	R
O	I	15	9	24	X
L	T	12	20	6	F
A	Q	1	17	18	R
R	U	18	21	13	M
I	B	9	2	11	K
Z	I	0	9	9	I
A	T	1	20	21	U
T	Q	20	17	11	K
I	U	9	21	4	D
O	B	15	2	17	Q
N	I	14	9	23	W

Here we wrote down the secret word repeatedly and added the codes of the letters of the secret word to the codes of the letters in the message. The decryption is done in a similar fashion using subtraction, instead of addition.

Another generalization of the Caesar's cypher is the substitution cypher, where instead of a shift in the alphabet, a fixed secret permutation of the alphabet letters is used, for example:

```
A B C D E F G H I J K L M N O P Q R S T U V W X Y Z
R G U N Q A W E V B P F D I Z T J Y C X M S O H K L
```

A permutation code may be broken using statistical analysis of the cyphertext. The idea of this attack is based on the fact that letters in the English language have noticeably different frequencies, with letter E being the most frequent letter, and Z being the least frequent. Here is the table of frequencies (in percent):

E	T	A	O	I	N	S	R	H	D	L	U	C
12	9.1	8.1	7.7	7.3	7	6.3	6.0	5.9	4.3	4.0	2.9	2.7

M	F	Y	W	G	P	B	V	K	X	Q	J	Z
2.6	2.3	2.1	2.1	2.0	1.8	1.5	1.1	.7	.17	.11	.10	.07

Counting frequencies of different letters in the cyphertext, we can identify several pairs of letters in the permutation, after which it is not too difficult to guess all other substitutions of letters.

The same idea is applicable to the Vigenère cyphers as well, only we have to use trial-and-error to determine the length of the secret codeword. If we make a guess that the secret codeword has length 5, then this would mean that 1st, 6th, 11th, 16th, etc., letters of the cyphertext are encrypted using the same substitution. Then separating this subsequence of letters, we can perform the frequency analysis on this subset of the cyphertext. If the resulting frequencies do not look like frequencies of letters in the English language, this would indicate that we guessed the length of the secret codeword incorrectly. If we guessed correctly, we should be able to figure out the value of the shift for this group of letters. Repeating the procedure for each group, we can uncover the secret codeword.

The weakness of the Vigenère method is in the limited size of the secret key, whereas modern requirement for cryptography is to have an ability of encryption of continuous streams of data. On the other hand, should the two parties have access to a common secret stream of random bits, they can use it to mask the stream of data by using addition of bit values mod 2, and such encryption is absolutely not breakable:

Random secret

stream: 0 0 1 0 1 1 0 0 0 1 0 1 0 1 1 0 1 1 0 1 0 1 0 0 1 1 1 0 1

Data

stream: 0 0 1 1 0 0 1 1 0 0 1 1 0 0 0 0 1 1 1 1 0 0 0 0 1 1 0 0 1

Encrypted

stream: 0 0 0 1 1 1 1 1 0 1 1 0 0 1 1 0 0 0 1 0 0 1 0 0 0 0 0 1 0 0

In this set-up the encrypted stream is not distinguishable from a random stream of data, and there can be no way to derive any information about the data stream if the secret key is not known. The recipient of the encrypted message can easily recover the data stream by adding the same secret stream.

The same secret key should never be used twice, doing so will compromise security of encryption. For this reason, this method is called *one-time pad.*

The conclusion of this discussion is that we can organize a provably secure encryption of a stream of data, should the two parties have access to a common secret random stream of bits. This is where quantum methods enter the game.

Here we outline two quantum key distribution protocols.

We begin with a protocol proposed by Bennett and Brassard in 1984 (BB84). In this scheme Alice will send to Bob a stream of linearly polarized photons with polarizations at angles $0°, 45°, 90°$ or $135°$. For each photon Alice chooses the angle of polarization at random.

Bob will perform the measurements of the photons he receives using one of the two polarizing filters: the first filter that passes photons with a vertical ($90°$) polarization and reflects photons with the horizontal polarization, and the second filter that passes photons polarized at $45°$ degrees and reflects photons polarized at $135°$.

Each filter corresponds to a basis in the 2-dimensional space of 1-qubits: the first filter corresponds to the standard basis $|0\rangle, |1\rangle$, and the second filter corresponds to the diagonal basis $\frac{1}{\sqrt{2}}|0\rangle + \frac{1}{\sqrt{2}}|1\rangle$, $-\frac{1}{\sqrt{2}}|0\rangle + \frac{1}{\sqrt{2}}|1\rangle$.

To write vectors in a more graphical way, we will use notations

$$\rightarrow = |0\rangle = \text{``0''}, \quad \uparrow = |1\rangle = \text{``1''},$$

$$\nearrow = \frac{1}{\sqrt{2}}|0\rangle + \frac{1}{\sqrt{2}}|1\rangle = \text{``0''}, \quad \nwarrow = -\frac{1}{\sqrt{2}}|0\rangle + \frac{1}{\sqrt{2}}|1\rangle = \text{``1''}.$$

Alice knows the state of each photon she generated, but Bob doesn't. Because of this he can only guess what is the appropriate filter for each photon. If Bob makes a right guess, he will determine correctly the polarization state of the photon and will get the corresponding bit value. If he uses a wrong filter to measure the photon, the outcome of the measurement will be random, and Bob will register either "0" or "1", each with 50% probability.

Once the measurements are done, Alice and Bob use unsecure (public) channel to reveal which basis was used by Alice to generate each photon (but not their states), and which basis was used by Bob for each measurement. They discard the bits in positions where the two bases did not match.

In the following example, the first row shows the states of photons sent by Alice, the second row is a sequence of bases used by Bob for measurements (S: standard, D: diagonal), the third row shows the outcomes of Bob's measurements, and the last row is the generated key, with discarded bits in positions where Alice's and Bob's bases did not match.

→	↗	↖	↑	↑	↖	→	↖	↗	↑	→	↑	→	↑	↖	→	→	↖
S	D	S	S	D	S	S	D	S	D	S	D	S	S	D	D	S	D
0	0	1	1	0	1	0	1	1	1	0	1	0	1	1	0	0	1
0	0		1			0	1			0		0	1	1		0	1

This results in the key 00101001101 which can be used to encrypt data exchanged between Alice and Bob using bitwise addition mod 2 of data bits and secret key bits.

The security of this scheme is based on the fact that in quantum mechanics measurement does not give full information about the state and alters the quantum state. This means that the attacker is not able to intercept Alice's photon, read its state and then send an identical photon to Bob. In order to verify the integrity of their transmission, Alice and Bob should publicly reveal a certain percentage of the common key. The revealed bits are of course discarded from the key. A man-in-the-middle attack will be immediately detected, since in such case the revealed bits will not match.

A potential weakness of BB84 is that lasers often send groups of photons in an identical state, instead of a single photon at a time. This opens the door to man-in-the-middle attack, where the attacker will intercept some of the photons in the group. If we have several photons in identical states, we can perform multiple measurements and deduce more information about the quantum state.

Provided that Alice is able to send to Bob a single photon at a time, BB84 scheme becomes provably unbreakable.

Another quantum protocol for the key distribution was proposed by Ekert in 1991. The E91 protocol exploits the phenomenon of entanglement. In this method we need to have a source generating pairs of entangled photons in identical states $\frac{1}{\sqrt{2}} |00\rangle + \frac{1}{\sqrt{2}} |11\rangle$. The first photon in each pair is sent to Alice, and the second photon is sent to Bob. When Alice and Bob perform measurements on the photons they receive, they may obtain either "0" or "1" with probability 50%, however there will be a perfect correlation between their measurements: they will either both observe "0" or both get "1". The resulting binary sequence will be random and may be used as a one-time pad.

A possible man-in-the middle attack can be attempted by the adversary Eve who may replace the original source of entangled photons by two beams of unentangled photons in the states $|00\rangle$ or $|11\rangle$. This would allow Eve to know the results of Alice's and Bob's observations and thus the secret key. However Alice and Bob have a way to verify that the photons they receive are indeed entangled and detect Eve's attack. We shall discuss the details of this in the next chapter.

Photons of an entangled pair were sent in opposite directions over a fiber optic cable achieving physical separation of hundreds of kilometers and their entanglement properties were successfully verified. In 2017, the Chinese satellite Micius sent photons in entangled pairs to two ground stations at a distance of 1200 km from each other.

4   Linear Transformations

Consider the following problem: we take a vector $\mathbf{v} = \begin{pmatrix} 3 \\ 2 \end{pmatrix}$ in the XY-plane and rotate it counterclockwise by 20°. What are the coordinates of the resulting vector $R_{20^\circ}(\mathbf{v})$?

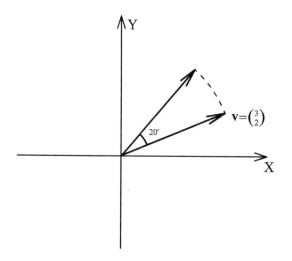

One can try finding a direct solution based on the geometry of triangles, but it is fairly complicated. There is a better approach based on the theoretical analysis of the properties of the rotation transformation. If we have two vectors $\mathbf{v}$ and $\mathbf{w}$ then it does not matter whether we add them and then rotate the sum by angle α or we first rotate each vector and then take the sum – the final result is going to be the same. This property of rotations can be expressed algebraically in the following way:

$$R_\alpha(\mathbf{v} + \mathbf{w}) = R_\alpha(\mathbf{v}) + R_\alpha(\mathbf{w}).$$

The same applies to the operation of multiplication of a vector by a number: for any real number c,

$$R_\alpha(c\mathbf{v}) = cR_\alpha(\mathbf{v}).$$

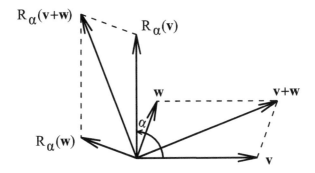

We observe that it is fairly easy to compute the result of the rotation transformation when it is applied to the basis vectors $\begin{pmatrix} 1 \\ 0 \end{pmatrix}$ and $\begin{pmatrix} 0 \\ 1 \end{pmatrix}$:

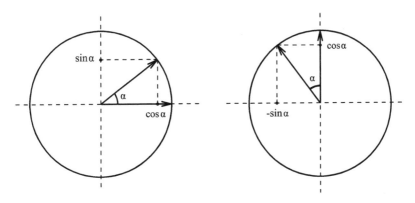

$$R_\alpha \begin{pmatrix} 1 \\ 0 \end{pmatrix} = \begin{pmatrix} \cos\alpha \\ \sin\alpha \end{pmatrix}, \quad R_\alpha \begin{pmatrix} 0 \\ 1 \end{pmatrix} = \begin{pmatrix} -\sin\alpha \\ \cos\alpha \end{pmatrix}.$$

Now we can compute the result of the rotation transformation applied to vector $\mathbf{v} = \begin{pmatrix} 3 \\ 2 \end{pmatrix}$ by writing it as

$$\begin{pmatrix} 3 \\ 2 \end{pmatrix} = 3 \begin{pmatrix} 1 \\ 0 \end{pmatrix} + 2 \begin{pmatrix} 0 \\ 1 \end{pmatrix},$$

and obtain

$$R_{20°}\begin{pmatrix}3\\2\end{pmatrix} = 3R_{20°}\begin{pmatrix}1\\0\end{pmatrix} + 2R_{20°}\begin{pmatrix}0\\1\end{pmatrix}$$

$$= 3\begin{pmatrix}\cos 20°\\\sin 20°\end{pmatrix} + 2\begin{pmatrix}-\sin 20°\\\cos 20°\end{pmatrix} \approx \begin{pmatrix}2.135\\2.905\end{pmatrix}.$$

We now want to adapt the idea of this computation to a much more general setting.

First, a *vector space* is a set of vectors in which we have an operation of addition of vectors and an operation of multiplication of vectors by numbers, satisfying a certain list of properties. We are not going to give this list here, but just say that these are all the natural properties that we would expect, like $c(\mathbf{v} + \mathbf{w}) = c\mathbf{v} + c\mathbf{w}$ for any two vectors $\mathbf{v}, \mathbf{w}$ and any number c.

Examples of vector spaces are:

- The set of plane vectors,

- The set of vectors in a 3D space,

- The set of n-qubits.

There is an algebraic construction that unifies all of the above examples. The space $\mathbb{R}^N$ is defined as a set of N-component arrays of real numbers. Addition and multiplication by real numbers are performed component-wise:

$$\begin{pmatrix}a_1\\a_2\\a_3\\\ldots\\a_N\end{pmatrix} + \begin{pmatrix}b_1\\b_2\\b_3\\\ldots\\b_N\end{pmatrix} = \begin{pmatrix}a_1+b_1\\a_2+b_2\\a_3+b_3\\\ldots\\a_N+b_N\end{pmatrix} , \quad c\begin{pmatrix}a_1\\a_2\\a_3\\\ldots\\a_N\end{pmatrix} = \begin{pmatrix}ca_1\\ca_2\\ca_3\\\ldots\\ca_N\end{pmatrix}.$$

The philosophy proposed by René Descartes is to have two models for the space of plane vectors: a geometric model, where vectors are the directed segments on a plane, and an algebraic one, where vectors are given by pairs of numbers. The advantage of the geometric model is that we can visualize it, while the algebraic model is better for doing calculations. Some problems are easier to solve with the geometric model, while others with the algebraic.

The space $\mathbb{R}^N$ has a basis $\{\mathbf{e}_1, \mathbf{e}_2, \ldots, \mathbf{e}_N\}$, where

$$
\mathbf{e}_1 = \begin{pmatrix} 1 \\ 0 \\ 0 \\ \cdots \\ 0 \end{pmatrix}, \quad
\mathbf{e}_2 = \begin{pmatrix} 0 \\ 1 \\ 0 \\ \cdots \\ 0 \end{pmatrix}, \quad \ldots, \quad
\mathbf{e}_N = \begin{pmatrix} 0 \\ 0 \\ 0 \\ \cdots \\ 1 \end{pmatrix}.
$$

Every vector in $\mathbb{R}^N$ may be expanded in a linear combination of the basis vectors:

$$
\begin{pmatrix} a_1 \\ a_2 \\ a_3 \\ \cdots \\ a_N \end{pmatrix} = \begin{pmatrix} a_1 \\ 0 \\ 0 \\ \cdots \\ 0 \end{pmatrix} + \begin{pmatrix} 0 \\ a_2 \\ 0 \\ \cdots \\ 0 \end{pmatrix} + \ldots + \begin{pmatrix} 0 \\ 0 \\ 0 \\ \cdots \\ a_N \end{pmatrix}
$$

$$
= a_1 \mathbf{e}_1 + a_2 \mathbf{e}_2 + a_3 \mathbf{e}_3 + \ldots + a_N \mathbf{e}_N.
$$

Comparing this with the expression for the 2-qubit,

$$
a_0 \left|00\right\rangle + a_1 \left|01\right\rangle + a_2 \left|10\right\rangle + a_3 \left|11\right\rangle,
$$

we see that the basis of the space of 2-qubits consists of the 4 pure state vectors $\{\left|00\right\rangle, \left|01\right\rangle, \left|10\right\rangle, \left|11\right\rangle\}$. This clearly generalizes to an arbitrary n-qubit space.

The dimension of a vector space is the number of vectors in its basis. We see that the dimension of $\mathbb{R}^N$ is N, while the dimension of the space of n-qubits is 2^n.

In addition to finite-dimensional spaces, such as those mentioned above, there are also infinite-dimensional vector spaces. An example of an infinite-dimensional vector space is the set of polynomials in a variable X. Just as with plane vectors, there is an operation of addition of two polynomials, and an operation of multiplication of a polynomial by a number. What is a basis in the space of polynomials? A polynomial can be written as

$$a_0 + a_1 X + a_2 X^2 + \ldots + a_n X^n.$$

We see that the coefficients $a_0, a_1, a_2, \ldots$ can be interpreted as the coordinates of a vector and $\{X^0, X^1, X^2, \ldots\}$ as a basis in the space of polynomials. Since there are infinitely many powers of X in the basis, the space of polynomials is infinite-dimensional.

Definition. A transformation T of a vector space V is called *linear* if it satisfies the following two properties:

$$T(\mathbf{v} + \mathbf{w}) = T(\mathbf{v}) + T(\mathbf{w}), \quad \text{for all } \mathbf{v}, \mathbf{w} \text{ in } V,$$

$$T(c\mathbf{v}) = c\, T(\mathbf{v}), \quad \text{for any number } c \text{ and all } \mathbf{v} \text{ in } V.$$

Rotation of a plane is an example of a linear transformation. Other examples are reflection of a plane in a line passing through the origin and dilation of a plane by a fixed factor.

A key feature of a linear transformation is that it is completely determined by what it does to the basis vectors. Suppose T is a linear transformation of $\mathbb{R}^N$, and we know what $T(\mathbf{e}_1), \ldots, T(\mathbf{e}_N)$ are. Then an arbitrary vector

$$\mathbf{v} = a_1 \mathbf{e}_1 + a_2 \mathbf{e}_2 + \ldots + a_N \mathbf{e}_N$$

gets transformed into

$$T(\mathbf{v}) = T(a_1\mathbf{e}_1 + a_2\mathbf{e}_2 + \ldots + a_N\mathbf{e}_N) = T(a_1\mathbf{e}_1) + T(a_2\mathbf{e}_2) + \ldots + T(a_N\mathbf{e}_N)$$

$$= a_1 T(\mathbf{e}_1) + a_2 T(\mathbf{e}_2) + \ldots + a_N T(\mathbf{e}_N).$$

This is the method we used to compute the image of the vector $\begin{pmatrix} 3 \\ 2 \end{pmatrix}$ under the rotation transformation.

Another example is for those who are familiar with calculus. Let us consider a transformation D of the space of polynomials, where each polynomial $f(X)$ is transformed into its derivative, $D(f) = f'$. Then the linear properties of D,

$$D(f + g) = D(f) + D(g), \quad D(cf) = cD(f),$$

are just the well-known sum rule and the constant multiple rule for the derivative. Hence differentiation is a linear transformation. We also learn in calculus that $D(X^k) = kX^{k-1}$. We point out that the computation of the derivative of an arbitrary polynomial is done precisely according to our approach – a polynomial is expanded into a combination of the basis vectors, and then the derivative is computed using its linear properties, for example:

$$D(X^5 + 3X^2 - 4X + 1) = D(X^5) + 3D(X^2) - 4D(X) + D(1)$$
$$= 5X^4 + 6X - 4.$$

Let us now apply the theory of linear transformations to quantum cryptography. Recall that for the key distribution scheme E91, we have a source producing pairs of entangled photons in the joint state $\frac{1}{\sqrt{2}} |00\rangle + \frac{1}{\sqrt{2}} |11\rangle$. One photon in each pair is sent to Alice, while the second photon is sent to Bob. When they perform the measurement, they will each observe 0 or 1 with probability 50%, but the outcome of each measurement will be the same for Alice and Bob.

Imagine that Eve tries to attack this scheme, and manages to replace the stream of entangled pairs with a stream of unentangled pairs, sending at random pairs in the states $|00\rangle$ or $|11\rangle$. In this case Alice and Bob will still have a 100% agreement in their observations, and half of their measurements will be zeros and half will be ones. Since these states are generated by Eve, Eve will know the outcomes of observations by Alice and Bob, and will thus know the generated key.

How can Alice and Bob detect this attack? Suppose Alice and Bob both rotate their polarization filters by the same angle α. How can we predict the outcomes of their observations? Mathematically, this is equivalent to applying a rotation transformation to each qubit and then performing the measurement. Quantum states of the photons will transform in the following way:

$$|0\rangle \mapsto \cos\alpha\,|0\rangle + \sin\alpha\,|1\rangle$$
$$|1\rangle \mapsto -\sin\alpha\,|0\rangle + \cos\alpha\,|1\rangle$$

Since the same transformation is applied to the first and the second photon in each pair, we get

$$\frac{1}{\sqrt{2}}\,|00\rangle + \frac{1}{\sqrt{2}}\,|11\rangle \mapsto$$

$$\frac{1}{\sqrt{2}}\,(\cos\alpha\,|0\rangle + \sin\alpha\,|1\rangle)(\cos\alpha\,|0\rangle + \sin\alpha\,|1\rangle)$$

$$+\frac{1}{\sqrt{2}}\,(-\sin\alpha\,|0\rangle + \cos\alpha\,|1\rangle)(-\sin\alpha\,|0\rangle + \cos\alpha\,|1\rangle)$$

$$=\frac{1}{\sqrt{2}}\,\left(\cos^2\alpha\,|00\rangle + \sin\alpha\cos\alpha\,|01\rangle + \sin\alpha\cos\alpha\,|10\rangle + \sin^2\alpha\,|11\rangle\right)$$

$$+\frac{1}{\sqrt{2}}\,\left(\sin^2\alpha\,|00\rangle - \sin\alpha\cos\alpha\,|01\rangle - \sin\alpha\cos\alpha\,|10\rangle + \cos^2\alpha\,|11\rangle\right)$$

$$=\frac{1}{\sqrt{2}}\,\left(\cos^2\alpha + \sin^2\alpha\right)|00\rangle + \frac{1}{\sqrt{2}}\,\left(\cos^2\alpha + \sin^2\alpha\right)|11\rangle$$

$$=\frac{1}{\sqrt{2}}\,|00\rangle + \frac{1}{\sqrt{2}}\,|11\rangle.$$

We arrive at an unexpected result – after performing the rotation, this entangled state did not change! This means that after rotating their polarization filters to the same new axis, there will be no change in the statistics of observations – Alice and Bob will still be observing zeros and ones with a 100% correlation.

Now let us see what will happen to the unentangled states $|00\rangle$ and $|11\rangle$ that Eve might send:

$$|00\rangle \mapsto (\cos\alpha\,|0\rangle + \sin\alpha\,|1\rangle)(\cos\alpha\,|0\rangle + \sin\alpha\,|1\rangle)$$
$$= \cos^2\alpha\,|00\rangle + \sin\alpha\cos\alpha\,|01\rangle + \sin\alpha\cos\alpha\,|10\rangle + \sin^2\alpha\,|11\rangle.$$

We see that in this case, it is possible that in some observations Alice will get 0, while Bob will observe 1 and vice versa. We can easily determine the probability of this, as a function of angle α.

In order to verify the integrity of the channel, Alice and Bob should be measuring a part of their stream with both of their polarization filters rotated by the same angle α. They should exchange the results of their measurements over a non-secure communication channel. A 100% correlation will indicate the absence of the attack. Naturally, the bits used for this test should not be used for the secret key generation.

We conclude this chapter explaining the idea of *parallelism* in quantum computing. We present it here in a crude, simplified form and will give more details in a later chapter. As we mentioned in the chapter on quantum mechanics, a quantum algorithm is a linear transformation of the space of n-qubits. Suppose we are interested in the values of a function $f(x)$ for $x = 0, 1, 2, \ldots, 2^n - 1$, and assume that the values of f are n-bit integers. We may define a linear transformation F of the space of n-qubits, which transforms a basis vector $|k\rangle$ into another basis vector $|f(k)\rangle$. Then the initial state

$$\sum_{k=0}^{2^n-1} a_k \, |k\rangle$$

will be transformed by quantum algorithm F into the state

$$\sum_{k=0}^{2^n-1} a_k \, |f(k)\rangle \, .$$

We see that in a single run of a quantum algorithm, we are able to get a state which incorporates all values of the function f. Coupled with the observation that a loop with 2^n iterations cannot be executed on a classical computer in a reasonable time (for example, the age of the Universe) for rather small values of n, like $n = 100$, we see that quantum computing allows massive parallelism, which cannot be matched by a classical computer.

5 The Matrix

In this chapter we are going to introduce matrix algebra – the technique of computations with the linear transformations.

We begin with the dot product in $\mathbb{R}^N$. The dot product of two vectors in $\mathbb{R}^N$ is a number defined in the following way:

$$\begin{pmatrix} a_1 \\ a_2 \\ \ldots \\ a_N \end{pmatrix} \cdot \begin{pmatrix} b_1 \\ b_2 \\ \ldots \\ b_N \end{pmatrix} = a_1 b_1 + a_2 b_2 + \ldots + a_N b_N.$$

The dot product is *bilinear*:

$$\mathbf{v} \cdot (\mathbf{u} + \mathbf{w}) = \mathbf{v} \cdot \mathbf{u} + \mathbf{v} \cdot \mathbf{w}, \quad (\mathbf{v} + \mathbf{u}) \cdot \mathbf{w} = \mathbf{v} \cdot \mathbf{w} + \mathbf{u} \cdot \mathbf{w},$$
$$\mathbf{v} \cdot (c\mathbf{w}) = (c\mathbf{v}) \cdot \mathbf{w} = c(\mathbf{v} \cdot \mathbf{w})$$

and *symmetric*: $\mathbf{v} \cdot \mathbf{w} = \mathbf{w} \cdot \mathbf{v}$.

An important property of the dot product is that the length of a vector $\mathbf{v}$ is equal to $\sqrt{\mathbf{v} \cdot \mathbf{v}}$. The length of a vector $\mathbf{v}$, also called the *norm* of $\mathbf{v}$, is denoted by $|\mathbf{v}|$.

Indeed, for a plane vector $\mathbf{v} = \begin{pmatrix} x \\ y \end{pmatrix}$ this is the Pythagoras Theorem:

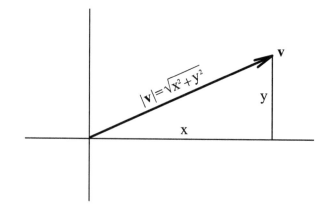

For a 3D vector $\mathbf{w} = \begin{pmatrix} x \\ y \\ z \end{pmatrix}$, first apply the Pythagoras Theorem

to the plane vector $\mathbf{v} = \begin{pmatrix} x \\ y \\ 0 \end{pmatrix}$, getting $|\mathbf{v}|^2 = x^2 + y^2$. Next we apply

the Pythagoras Theorem again to the pair of perpendicular vectors

$\mathbf{v} = \begin{pmatrix} x \\ y \\ 0 \end{pmatrix}$ and $\mathbf{u} = \begin{pmatrix} 0 \\ 0 \\ z \end{pmatrix}$ and we get $|\mathbf{w}|^2 = |\mathbf{v}|^2 + |\mathbf{u}|^2 = x^2 + y^2 + z^2$.

To establish the formula $\mathbf{v} \cdot \mathbf{v} = |\mathbf{v}|^2$ in $\mathbb{R}^N$ we need to apply the Pythagoras Theorem in a similar fashion $N - 1$ times.

Theorem. Let $\mathbf{u}, \mathbf{v}$ be two vectors in $\mathbb{R}^N$ with the angle α between them. Then $\mathbf{u} \cdot \mathbf{v} = |\mathbf{u}||\mathbf{v}| \cos \alpha$.

Proof. Consider a triangle formed by the vectors $\mathbf{u}$, $\mathbf{v}$ and $\mathbf{u} - \mathbf{v}$.

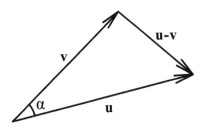

Using the properties of the dot product,

$$|\mathbf{u} - \mathbf{v}|^2 = (\mathbf{u} - \mathbf{v}) \cdot (\mathbf{u} - \mathbf{v}) = \mathbf{u} \cdot \mathbf{u} - \mathbf{u} \cdot \mathbf{v} - \mathbf{v} \cdot \mathbf{u} + \mathbf{v} \cdot \mathbf{v} = |\mathbf{u}|^2 + |\mathbf{v}|^2 - 2\mathbf{u} \cdot \mathbf{v}.$$

By Cosine Theorem, however,

$$|\mathbf{u} - \mathbf{v}|^2 = |\mathbf{u}|^2 + |\mathbf{v}|^2 - 2|\mathbf{u}||\mathbf{v}| \cos \alpha.$$

Comparing the two formulas we obtain the claim of the Theorem.

Corollary. Two vectors in $\mathbb{R}^N$ are perpendicular to each other if and only if their dot product is zero.

Definition. The *matrix* of a linear transformation T of the vector space $\mathbb{R}^N$ is a square $N \times N$ table of numbers, formed by the vectors $T(\mathbf{e}_1), T(\mathbf{e}_2), \ldots, T(\mathbf{e}_N)$, placed as columns.

Example: Let T be the rotation of the plane $30°$ counterclockwise. Then

$$T(\mathbf{e}_1) = \begin{pmatrix} \cos 30° \\ \sin 30° \end{pmatrix} = \begin{pmatrix} \sqrt{3}/2 \\ 1/2 \end{pmatrix}, \quad T(\mathbf{e}_2) = \begin{pmatrix} -\sin 30° \\ \cos 30° \end{pmatrix} = \begin{pmatrix} -1/2 \\ \sqrt{3}/2 \end{pmatrix},$$

and the matrix of T is

$$\begin{pmatrix} \sqrt{3}/2 & -1/2 \\ 1/2 & \sqrt{3}/2 \end{pmatrix}.$$

In general, the matrix of a rotation transformation of a plane in angle α counterclockwise, is:

$$R_\alpha = \begin{pmatrix} \cos \alpha & -\sin \alpha \\ \sin \alpha & \cos \alpha \end{pmatrix}.$$

As we saw in the previous chapter, a linear transformation is determined by the images of the basis vectors. Hence, the matrix of a linear transformation encodes complete information about the transformation.

Next we define the operation of multiplication of an $N \times N$ matrix by a vector from $\mathbb{R}^N$.

Definition. The product $A\mathbf{v}$ of an $N \times N$ matrix A with a vector $\mathbf{v}$ in $\mathbb{R}^N$ is a vector, whose k-th component is the dot product of k-th row of A with the vector $\mathbf{v}$.

Theorem. Let T be a linear transformation of $\mathbb{R}^N$ with matrix A. Then the result of applying T to a vector $\mathbf{v}$ is equal to the product $A\mathbf{v}$.

Instead of giving a formal proof, let us consider an example. Let T be a linear transformation of $\mathbb{R}^2$ with matrix

$$A = \begin{pmatrix} 1 & 2 \\ 3 & 4 \end{pmatrix}$$

and let $\mathbf{v} = \begin{pmatrix} 5 \\ 6 \end{pmatrix}$. Then the product of A with $\mathbf{v}$ is:

$$Av = \begin{pmatrix} 1 & 2 \\ 3 & 4 \end{pmatrix} \begin{pmatrix} 5 \\ 6 \end{pmatrix} = \begin{pmatrix} 1 \times 5 + 2 \times 6 \\ 3 \times 5 + 4 \times 6 \end{pmatrix} = \begin{pmatrix} 17 \\ 39 \end{pmatrix},$$

while the image of $\mathbf{v}$ under the transformation T is:

$$T(\mathbf{v}) = T(5\mathbf{e}_1 + 6\mathbf{e}_2) = 5T(\mathbf{e}_1) + 6T(\mathbf{e}_2)$$
$$= 5 \begin{pmatrix} 1 \\ 3 \end{pmatrix} + 6 \begin{pmatrix} 2 \\ 4 \end{pmatrix} = \begin{pmatrix} 5 \times 1 + 6 \times 2 \\ 5 \times 3 + 6 \times 4 \end{pmatrix} = \begin{pmatrix} 17 \\ 39 \end{pmatrix}.$$

We can see that this computation remains valid for all $N \times N$ matrices and vectors in $\mathbb{R}^N$.

Exercise. Let T be a reflection of the plane with respect to the line $y = 2x$. Find the matrix of T.

Solution. To form the matrix of T we need to find $T(\mathbf{e}_1)$ and $T(\mathbf{e}_2)$. We note that the vector $\mathbf{v}_1 = \begin{pmatrix} 1 \\ 2 \end{pmatrix}$ is on the line $y = 2x$. Then the vector $\mathbf{v}_2 = \begin{pmatrix} 2 \\ -1 \end{pmatrix}$ is perpendicular to the line (the dot product of these two vectors is zero). Decompose vector $\mathbf{e}_1$ as a linear combination of $\mathbf{v}_1$ and $\mathbf{v}_2$: $\mathbf{e}_1 = c_1\mathbf{v}_1 + c_2\mathbf{v}_2$:

$$\begin{pmatrix} 1 \\ 0 \end{pmatrix} = c_1 \begin{pmatrix} 1 \\ 2 \end{pmatrix} + c_2 \begin{pmatrix} 2 \\ -1 \end{pmatrix}.$$

Looking at the second components of vectors in this equation, we see that $c_2 = 2c_1$. Making this substitution into the equation for the first components, we can easily find c_1. We get $c_1 = 1/5$, $c_2 = 2/5$, hence $\mathbf{e}_1 = 1/5\mathbf{v}_1 + 2/5\mathbf{v}_2$. Since $\mathbf{v}_1$ is on the line of reflection, we get that $T(\mathbf{v}_1) = \mathbf{v}_1$. Vector $\mathbf{v}_2$ is perpendicular to the line of reflection, so its mirror image is $T(\mathbf{v}_2) = -\mathbf{v}_2$. Combining this with the expression for $\mathbf{e}_1$, we get

$$T(\mathbf{e}_1) = 1/5T(\mathbf{v}_1) + 2/5T(\mathbf{v}_2) = 1/5 \begin{pmatrix} 1 \\ 2 \end{pmatrix} - 2/5 \begin{pmatrix} 2 \\ -1 \end{pmatrix} = \begin{pmatrix} -3/5 \\ 4/5 \end{pmatrix}.$$

In a similar way we find $T(e_2) = \begin{pmatrix} 4/5 \\ 3/5 \end{pmatrix}$. Placing $T(e_1)$ and $T(e_2)$ as columns, we obtain the matrix of T:

$$\begin{pmatrix} -3/5 & 4/5 \\ 4/5 & 3/5 \end{pmatrix}.$$

In classical computing, complicated problems are rarely solved in one step. Typically we perform a sequence of operations to arrive at the answer. Likewise, a quantum algorithm is designed not as a single linear transformation, but as a composition of several simpler linear transformations.

We need to figure out how to calculate the matrix of a composition $T \circ S$ of linear transformations if we know matrices for T and S.

First we need to discuss a convention on the order of operations. It is traditional in mathematics to write the argument of a function $f(x)$ or of a linear transformation $T(\mathbf{v})$ on the right. The composition $T \circ S$ when applied to a vector $\mathbf{v}$ gives $T \circ S(\mathbf{v}) = T(S(\mathbf{v}))$, which means that in $T \circ S$ the factor that appears on the right is applied first.

Let T and S be two linear transformations of $\mathbb{R}^N$ with matrices A and B respectively. We wish to calculate the matrix C of the composition $T \circ S$. Recall that k-th column of C is the image of e_k under the transformation $T \circ S$: $T \circ S(e_k) = T(S(e_k))$. However, $S(e_k)$ is just the k-th column of matrix B. We conclude that the k-th column of C is the product of A with the k-th column of B. We call matrix C, computed in this way, the *product* of A and B, $C = AB$.

For example,

$$\begin{pmatrix} 1 & 2 \\ 3 & 4 \end{pmatrix} \begin{pmatrix} 5 & 1 \\ 2 & 0 \end{pmatrix} = \begin{pmatrix} 1 \times 5 + 2 \times 2 & 1 \times 1 + 2 \times 0 \\ 3 \times 5 + 4 \times 2 & 3 \times 1 + 4 \times 0 \end{pmatrix} = \begin{pmatrix} 9 & 1 \\ 23 & 3 \end{pmatrix}.$$

We see here that the number in row m, column k of AB is the dot product of row m of A with the column k of B.

To summarize, the matrix of the composition of two linear transformations is the product of matrices of these transformations.

We can also define the sum of two $N \times N$ matrices in a component-wise way:

$$\begin{pmatrix} 1 & 2 \\ 3 & 4 \end{pmatrix} + \begin{pmatrix} 5 & 1 \\ 2 & 0 \end{pmatrix} = \begin{pmatrix} 1+5 & 2+1 \\ 3+2 & 4+0 \end{pmatrix} = \begin{pmatrix} 6 & 3 \\ 5 & 4 \end{pmatrix}.$$

Some algebraic properties of matrix operations are the same as for numbers: $A(B+C) = AB + AC$, $(A+B)C = AC + BC$, yet there is one important difference. Consider the following two matrices:

$$A = \begin{pmatrix} 1 & 1 \\ -1 & -1 \end{pmatrix}, \quad B = \begin{pmatrix} 1 & 1 \\ 1 & 1 \end{pmatrix}.$$

Compute the products AB and BA:

$$AB = \begin{pmatrix} 2 & 2 \\ -2 & -2 \end{pmatrix}, \quad BA = \begin{pmatrix} 0 & 0 \\ 0 & 0 \end{pmatrix}.$$

We conclude that multiplication of matrices is *non-commutative*: $AB \neq BA$ in general!

This example also shows that unlike numbers, the product of two non-zero matrices could be zero matrix.

Still, the product of matrices is *associative*: $(AB)C = A(BC)$. This follows from the fact that matrix multiplication corresponds to the composition of linear transformations, and for linear transformations we have equality $(T \circ S) \circ R = T \circ (S \circ R)$, since both sides applied to vector $\mathbf{v}$ will yield $T(S(R(\mathbf{v})))$.

We conclude this chapter with an application of linear algebra to trigonometry. Consider a composition of a rotation in angle α with a rotation in angle β. Clearly,

$$R_\alpha \circ R_\beta = R_{\alpha+\beta}.$$

Let us express this equality with rotation matrices:

$$\begin{pmatrix} \cos\alpha & -\sin\alpha \\ \sin\alpha & \cos\alpha \end{pmatrix} \begin{pmatrix} \cos\beta & -\sin\beta \\ \sin\beta & \cos\beta \end{pmatrix} = \begin{pmatrix} \cos(\alpha+\beta) & -\sin(\alpha+\beta) \\ \sin(\alpha+\beta) & \cos(\alpha+\beta) \end{pmatrix}.$$

The product of matrices in the left hand side gives:

$$\begin{pmatrix} \cos\alpha\cos\beta - \sin\alpha\sin\beta & -\cos\alpha\sin\beta - \sin\alpha\cos\beta \\ \sin\alpha\cos\beta + \cos\alpha\sin\beta & -\sin\alpha\sin\beta + \cos\alpha\cos\beta \end{pmatrix}.$$

Comparing this with the rotation matrix in the right hand side, we obtain the trigonometric identities:

$$\cos(\alpha + \beta) = \cos\alpha\cos\beta - \sin\alpha\sin\beta,$$
$$\sin(\alpha + \beta) = \sin\alpha\cos\beta + \cos\alpha\sin\beta.$$

This is the most economical proof of these important formulas.

6 Orthogonal Linear Transformations

A quantum algorithm is a linear transformation of the space of qubits, but it is a transformation of a special kind, it is an *orthogonal* linear transformation.

Definition. A linear transformation T of vector space $\mathbb{R}^N$ is called orthogonal if the images of basis vectors $T(\mathbf{e}_1), T(\mathbf{e}_2), \ldots, T(\mathbf{e}_N)$ are orthogonal to each other and all have unit length.

Definition. A set of N mutually orthogonal length 1 vectors in $\mathbb{R}^N$ is called an *orthonormal* basis of $\mathbb{R}^N$.

Matrix of an orthogonal transformation is called an orthogonal matrix.

Here is the main property of orthogonal matrices: dot product of any two different columns of an orthogonal matrix is zero; dot product of any column with itself is equal to one. This follows immediately from the definition of the matrix of a linear transformation.

Let us give several examples of orthogonal transformations:

1. Rotation of $\mathbb{R}^2$ by angle α. It is clear from geometry that the vectors $R_\alpha(\mathbf{e}_1)$ and $R_\alpha(\mathbf{e}_2)$ are both unit vectors and perpendicular to each other. Observe that the dot products of the columns of the matrix of R_α are as expected:

$$\begin{pmatrix} \cos\alpha & -\sin\alpha \\ \sin\alpha & \cos\alpha \end{pmatrix}.$$

2. Reflection of $\mathbb{R}^2$ with respect to a line passing through the origin. This transformation is orthogonal because mirror reflection preserves lengths of vectors, as well as the angles between vectors.

3. Rotation of $\mathbb{R}^3$ in angle α around some axis passing through the origin.

In fact, we can show that any orthogonal transformation of $\mathbb{R}^2$ is either a rotation or a reflection, as in examples 1-2 above. Let us sketch an argument. Let T be an orthogonal transformation of $\mathbb{R}^2$ and let $\mathbf{u}_1 = T(\mathbf{e}_1)$ and $\mathbf{u}_2 = T(\mathbf{e}_2)$. Since T is orthogonal, we know that $\mathbf{u}_1$ and $\mathbf{u}_2$ are unit vectors and orthogonal to each other. All unit vectors on a plane can be obtained from each other by rotations. Suppose $\mathbf{u}_1$ may be obtained by rotating $\mathbf{e}_1$ by angle α in the counterclockwise direction. Then

$$\mathbf{u}_1 = \begin{pmatrix} \cos \alpha \\ \sin \alpha \end{pmatrix}.$$

Since $\mathbf{u}_2$ is perpendicular to $\mathbf{u}_1$, there are just two possibilities for it – $\mathbf{u}_2$ is obtained from $\mathbf{u}_1$ by rotating $\mathbf{u}_1$ by angle $90°$ counterclockwise or clockwise.

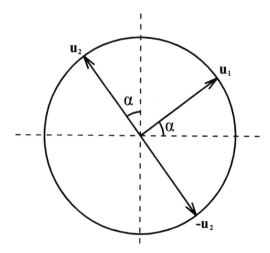

In the first case T is a rotation with a matrix written above, and in the second case T is a reflection with matrix

$$\begin{pmatrix} \cos \alpha & \sin \alpha \\ \sin \alpha & -\cos \alpha \end{pmatrix}.$$

Next we are going to establish several properties of orthogonal transformations.

Theorem. A linear transformation T is orthogonal if and only if it preserves the dot product:

$$T(\mathbf{u}) \cdot T(\mathbf{v}) = \mathbf{u} \cdot \mathbf{v} \quad \text{for all } \mathbf{u}, \mathbf{v} \text{ in } \mathbb{R}^N.$$

Proof. Assume that T is orthogonal. It follows from the definition that this claim is true for the basis vectors: $T(\mathbf{e}_i) \cdot T(\mathbf{e}_j) = \mathbf{e}_i \cdot \mathbf{e}_j$. To prove the claim for the arbitrary vectors $\mathbf{u}, \mathbf{v}$, expand them as linear combinations of basis vectors:

$$\mathbf{u} = \sum_{i=1}^{N} b_i \mathbf{e}_i, \quad \mathbf{v} = \sum_{j=1}^{N} c_j \mathbf{e}_j.$$

Then

$$T(\mathbf{u}) \cdot T(\mathbf{v}) = T\left(\sum_{i=1}^{N} b_i \mathbf{e}_i\right) \cdot T\left(\sum_{j=1}^{N} c_j \mathbf{e}_j\right)$$

$$= \left(\sum_{i=1}^{N} b_i T(\mathbf{e}_i)\right) \cdot \left(\sum_{j=1}^{N} c_j T(\mathbf{e}_j)\right) = \sum_{i=1}^{N}\sum_{j=1}^{N} b_i c_j \left(T(\mathbf{e}_i) \cdot T(\mathbf{e}_j)\right)$$

$$\sum_{i=1}^{N}\sum_{j=1}^{N} b_i c_j (\mathbf{e}_i \cdot \mathbf{e}_j) = \left(\sum_{i=1}^{N} b_i \mathbf{e}_i\right) \cdot \left(\sum_{j=1}^{N} c_j \mathbf{e}_j\right) = \mathbf{u} \cdot \mathbf{v},$$

and we conclude that an orthogonal transformation preserves the dot product.

Conversely, if a linear transformation T preserves the dot product, we get that $T(\mathbf{e}_i) \cdot T(\mathbf{e}_j) = \mathbf{e}_i \cdot \mathbf{e}_j$. Thus T transforms $\{\mathbf{e}_1, \ldots, \mathbf{e}_N\}$ into another orthonormal basis. Hence T is an orthogonal transformation.

A consequence of this Theorem is that orthogonal transformations preserve the length of a vector, as well as angles between vectors. This follows from the fact that lengths and angles can be expressed using dot products.

We also note that the dot product of two vectors can be expressed using lengths of vectors:

$$\mathbf{u} \cdot \mathbf{v} = \frac{1}{2}(|\mathbf{u} + \mathbf{v}|^2 - |\mathbf{u}|^2 - |\mathbf{v}|^2),$$

hence any linear transformation that preserves lengths of vectors also preserves dot products and must be orthogonal. Let us summarize:

Theorem. For a linear transformation T, the following four conditions are equivalent:

(1) T is an orthogonal transformation.
(2) T preserves dot products.
(3) T preserves lengths and angles.
(4) T preserves lengths of vectors.

In the above discussion we proved that $(1) \Leftrightarrow (2) \Rightarrow (3) \Rightarrow (4) \Rightarrow (2)$.

In the class of linear transformations there is one transformation that plays the role of number 1. The *identity transformation* I of a vector space is the transformation that does not change any vector: $I(\mathbf{v}) = \mathbf{v}$ for all vectors $\mathbf{v}$.

Consider the identity transformation of $\mathbb{R}^N$. Since $I(\mathbf{e}_1) = \mathbf{e}_1, \ldots,$ $I(\mathbf{e}_N) = \mathbf{e}_N$, the matrix of the identity transformation is diagonal with 1's along the diagonal. We denote this matrix also by I:

$$I = \begin{pmatrix} 1 & 0 & 0 & \cdots & 0 \\ 0 & 1 & 0 & \cdots & 0 \\ 0 & 0 & 1 & \cdots & 0 \\ \cdots & \cdots & \cdots & \cdots & \cdots \\ 0 & 0 & 0 & \cdots & 1 \end{pmatrix}.$$

The identity transformation has the following obvious properties with respect to composition: $T \circ I = I \circ T = T$ for all transformations T. As a consequence, we obtain the same properties of the identity matrix with respect to matrix multiplication: $AI = IA = A$.

We can also generalize for linear transformations the arithmetic operation of inversion:

Definition. A linear transformation S is called the inverse of T (denoted $S = T^{-1}$), if $T \circ S = S \circ T = I$.

Likewise, for matrices $B = A^{-1}$ if $AB = BA = I$.

Exercise. Verify that

$$\begin{pmatrix} 1 & 2 \\ 3 & 5 \end{pmatrix}^{-1} = \begin{pmatrix} -5 & 2 \\ 3 & -1 \end{pmatrix}.$$

For real numbers, the only non-invertible number is 0. For matrices, there are non-zero matrices that are not invertible.

Exercise. Show that the matrix

$$\begin{pmatrix} 1 & 1 \\ 1 & 1 \end{pmatrix}$$

does not have an inverse.

We state the following fact for $N \times N$ matrices without proof: If $AB = I$ then $BA = I$. It follows that the same is true for linear transformations of $\mathbb{R}^N$: if $T \circ S = I$ then $S \circ T = I$.

Exercise. Show that the above claim fails for linear transformations of infinite-dimensional vector spaces. Let D, S be linear transformations of the space of polynomials, such that $D(X^n) = nX^{n-1}$ and $S(X^n) = \frac{1}{n+1}X^{n+1}$. Show that $D \circ S = I$, but $S \circ D \neq I$.

Definition. The *transpose* of a matrix A is a matrix whose k-th row is the k-th column of A (notation: A^T).

Here is an example of the transpose:

$$\begin{pmatrix} 1 & 2 & 3 \\ 4 & 5 & 6 \\ 7 & 8 & 9 \end{pmatrix}^T = \begin{pmatrix} 1 & 4 & 7 \\ 2 & 5 & 8 \\ 3 & 6 & 9 \end{pmatrix}.$$

Theorem. Every orthogonal matrix A is invertible. Its inverse is its transpose: $A^{-1} = A^T$.

Proof. We need to show that $A^T A = I$. The entry in row i, column j of $A^T A$ is the dot product of row i of A^T and column j of A. But rows of A^T are columns of A, so this entry is the dot product of columns i and j of A. Since A is an orthogonal matrix, this product is 1 if $i = j$, and 0 otherwise, which yields the identity matrix.

7 Quantum Teleportation

Two objects in identical quantum states are physically not distinguishable from each other. The idea of quantum teleportation is in moving a *quantum state* from point A to point B, rather than moving *matter*. Quantum teleportation is real, and has been tested in a lab, however only for very small quantum systems. In theory, it is possible to teleport arbitrarily large quantum systems, but we do not yet possess technology to perform teleportation of macroscopic objects.

We would like to teleport a polarized photon from point A to point B. What we really mean here is that we want to teleport the state of the photon. A naive approach is to measure the state of the photon at location A, and to recreate a photon in an identical state at location B. This is not possible, since we cannot determine the unknown state of a photon with a single measurement, and such a measurement will invariably destroy the state.

The trick here is to teleport the state of the photon without ever finding out what the state actually is.

Imagine we are given a photon at location A in an unknown polarization state $a\,|0\rangle + b\,|1\rangle$ with $a^2 + b^2 = 1$.

In order to implement the teleportation, we should have done some preparation in advance. We should have generated an entangled pair in the state $\frac{1}{\sqrt{2}}\,|00\rangle + \frac{1}{\sqrt{2}}\,|11\rangle$ and send the first photon to location A, while the second photon of the pair to location B.

Now we get 3 photons, their joint state is a 3-qubit, which can be calculated as a tensor product:

$$(a\,|0\rangle + b\,|1\rangle)\left(\frac{1}{\sqrt{2}}\,|00\rangle + \frac{1}{\sqrt{2}}\,|11\rangle\right)$$

$$= \frac{a}{\sqrt{2}}\,|000\rangle + \frac{a}{\sqrt{2}}\,|011\rangle + \frac{b}{\sqrt{2}}\,|100\rangle + \frac{b}{\sqrt{2}}\,|111\rangle.$$

Next we are going to mix the states of the two photons at location A by performing the following orthogonal transformation of the space of 2-qubits:

$$|00\rangle \mapsto \frac{1}{2}\left(|00\rangle + |01\rangle + |10\rangle + |11\rangle\right),$$

$$|01\rangle \mapsto \frac{1}{2}\left(|00\rangle - |01\rangle + |10\rangle - |11\rangle\right),$$

$$|10\rangle \mapsto \frac{1}{2}\left(|00\rangle + |01\rangle - |10\rangle - |11\rangle\right),$$

$$|11\rangle \mapsto \frac{1}{2}\left(-|00\rangle + |01\rangle + |10\rangle - |11\rangle\right).$$

This transformation is orthogonal since it transforms each basis vector into a vector of length 1, and the images of different basis vectors have zero dot product with each other.

If we perform this transformation on the first two photons of the 3-qubit, we get the state

$$\frac{a}{2\sqrt{2}}\left(|000\rangle + |010\rangle + |100\rangle + |110\rangle\right)$$

$$+\frac{a}{2\sqrt{2}}\left(|001\rangle - |011\rangle + |101\rangle - |111\rangle\right)$$

$$+\frac{b}{2\sqrt{2}}\left(|000\rangle + |010\rangle - |100\rangle - |110\rangle\right)$$

$$+\frac{b}{2\sqrt{2}}\left(-|001\rangle + |011\rangle + |101\rangle - |111\rangle\right)$$

$$=\frac{a+b}{2\sqrt{2}}|000\rangle + \frac{a-b}{2\sqrt{2}}|001\rangle$$

$$+\frac{a+b}{2\sqrt{2}}|010\rangle + \frac{-a+b}{2\sqrt{2}}|011\rangle$$

$$+\frac{a-b}{2\sqrt{2}}|100\rangle + \frac{a+b}{2\sqrt{2}}|101\rangle$$

$$+\frac{a-b}{2\sqrt{2}}|110\rangle + \frac{-a-b}{2\sqrt{2}}|111\rangle.$$

Next we perform the measurement of the two photons at location A. There are four possible outcomes: 00, 01, 10, 11. Let us consider these four cases, and determine what will happen in each case to the state of the photon at location B.

Case 1. We observe 00 on the first two photons. In this case the 3-qubit will collapse to

$$\frac{a+b}{\sqrt{2}} |000\rangle + \frac{a-b}{\sqrt{2}} |001\rangle,$$

which factors in the tensor product

$$|00\rangle \left(\frac{a+b}{\sqrt{2}} |0\rangle + \frac{a-b}{\sqrt{2}} |1\rangle \right),$$

meaning that the state of the photon at location B becomes

$$\frac{a+b}{\sqrt{2}} |0\rangle + \frac{a-b}{\sqrt{2}} |1\rangle.$$

Case 2. We observe 01 on the first two photons. In this case the state of the photon at location B becomes

$$\frac{a+b}{\sqrt{2}} |0\rangle + \frac{-a+b}{\sqrt{2}} |1\rangle.$$

Case 3. We observe 10 on the first two photons. In this case the state of the photon at location B becomes

$$\frac{a-b}{\sqrt{2}} |0\rangle + \frac{a+b}{\sqrt{2}} |1\rangle.$$

Case 4. We observe 11 on the first two photons. In this case the state of the photon at location B becomes

$$\frac{a-b}{\sqrt{2}} |0\rangle + \frac{-a-b}{\sqrt{2}} |1\rangle.$$

Finally, we perform an orthogonal transformation of the qubit at location B. The type of this transformation will depend on the outcome of the observation at location A.

Case 1. If we observed 00 at A, we perform the following transformation of the photon at B:

$$|0\rangle \mapsto \frac{1}{\sqrt{2}} |0\rangle + \frac{1}{\sqrt{2}} |1\rangle,$$

$$|1\rangle \mapsto \frac{1}{\sqrt{2}} |0\rangle - \frac{1}{\sqrt{2}} |1\rangle.$$

Case 2. If we observed 01 at A, we perform the following transformation of the photon at B:

$$|0\rangle \mapsto \frac{1}{\sqrt{2}}|0\rangle + \frac{1}{\sqrt{2}}|1\rangle\,,$$

$$|1\rangle \mapsto -\frac{1}{\sqrt{2}}|0\rangle + \frac{1}{\sqrt{2}}|1\rangle\,.$$

Case 3. If we observed 10 at A, we perform the following transformation of the photon at B:

$$|0\rangle \mapsto \frac{1}{\sqrt{2}}|0\rangle - \frac{1}{\sqrt{2}}|1\rangle\,,$$

$$|1\rangle \mapsto \frac{1}{\sqrt{2}}|0\rangle + \frac{1}{\sqrt{2}}|1\rangle\,.$$

Case 4. If we observed 11 at A, we perform the following transformation of the photon at B:

$$|0\rangle \mapsto \frac{1}{\sqrt{2}}|0\rangle - \frac{1}{\sqrt{2}}|1\rangle\,,$$

$$|1\rangle \mapsto -\frac{1}{\sqrt{2}}|0\rangle - \frac{1}{\sqrt{2}}|1\rangle\,.$$

We can check that in all cases we obtain the same result at location B: $a|0\rangle + b|1\rangle$. Let us carry out the calculations for Case 1:

$$\frac{a+b}{\sqrt{2}}|0\rangle + \frac{a-b}{\sqrt{2}}|1\rangle \mapsto \frac{a+b}{2}|0\rangle + \frac{a+b}{2}|1\rangle + \frac{a-b}{2}|0\rangle - \frac{a-b}{2}|1\rangle$$

$$= a|0\rangle + b|1\rangle\,.$$

We see that the final result at B is the quantum state of the original qubit that was at A. We have succeeded in teleporting this state from A to B. Note that none of our transformations depended on the values of a or b. The transformation carried out on the last step depended on the observations obtained at A, but not on the values of the coefficients a, b, which remained undetermined throughout the whole process.

We point out that performing teleportation did require the transfer of matter – the two parts of the entangled 2-qubit had to be delivered to locations A and B. However this can be done in advance.

We also note that the speed of the quantum teleportation does not exceed the speed of light, so this procedure does not violate the principles of relativity theory. Indeed, the final transformation at location B depended on the outcome of observations at A. These outcomes must be transmitted from A to B, and the speed of that information transfer is bounded by the speed of light. We emphasize here that the information being transmitted is *classical*, it is just the value of two bits, so this information transfer may be done through an ordinary communication channel.

Finally, we remark that the original quantum state at location A has been destroyed. In fact, there is a theorem that states that an unknown quantum state cannot be duplicated, so teleporting a state from A to B necessitates the destruction of the original quantum state at A.

8   Group Theory

A group is a mathematical tool to study symmetry. The idea of group theory is to generalize the algebra of numbers to the algebra of symmetries. We can associate a group with any symmetric object, be that an object from the natural world, or an abstract mathematical construction.

Informally, a symmetry transformation of an object is a transformation that moves the points/elements of the object, yet preserves it as a whole. A good example to keep in mind is a rotation of a sphere. When we rotate a sphere its points move, but the sphere as a set of points, is preserved. A group is then the set of all symmetry transformations of a given object.

The concept of symmetry played a central role in the development of physics in the 20th century. Although the use of symmetry in physics goes back to Galileo, it was Einstein who made symmetry a cornerstone of physics. Einstein was able to build relativity theory from a single postulate about the geometry of space-time and its symmetry transformations.

There are some very general principles that are universally valid for symmetry transformations of any object:

1. Every symmetry transformation is invertible.

For example, if we rotate a sphere by angle α around a certain axis, to undo this transformation we can rotate it around the same axis by angle α in the opposite direction.

2. A composition of two symmetry transformations is a symmetry transformation.

We can build a symmetry transformation of a sphere by first rotating it around a certain axis by angle α, followed by a rotation around another axis by angle β. It is clear that the composition of these two transformations will still preserve the sphere. What is less obvious, is that such a composition may be expressed as a single rotation around some third axis.

3. There is always a trivial symmetry transformation that does not move any points. Such a transformation is called the identity

transformation.

With these principles in mind, we can present the formal definition of a group.

Definition. A group G is a set with a distinguished element e, called the identity, an operation of inversion of elements, $x \mapsto x^{-1}$ and an operation of multiplication of pairs of elements, $x \cdot y$, which satisfy the following properties for all x, y, z in G:

(1) $x \cdot e = e \cdot x = x$ [axiom of the identity],

(2) $x \cdot x^{-1} = x^{-1} \cdot x = e$ [axiom of the inverse],

(3) $(x \cdot y) \cdot z = x \cdot (y \cdot z)$ [associative law].

We can immediately see that axioms of the identity and the inverse hold for all symmetry transformations. It turns out that the associative law also holds universally for all compositions of transformations (the argument we used to prove the associative law for the compositions of linear transformations is applicable in a more general setting).

Just as with linear transformations, for the composition $f \cdot g$ of symmetry transformations, the factor on the right is applied first.

First examples of groups originate in the arithmetic.

Consider the set $\mathbb{Q}^*$ of non-zero rational numbers. This set is closed under the operations of multiplication and inversion. The identity element is the number 1. All three axioms of a group will hold, making $\mathbb{Q}^*$ a group.

Another possibility is to take the set $\mathbb{R}^*$ of non-zero real numbers. This will give us the second example of a group.

We can consider groups with operations other than multiplication. What is crucial is the properties of operations, and not their names or notations. Let us translate the definition of a group into a different language, where the group operation is addition, rather than multiplication. With addition, the axiom of identity will read: $x + e = e + x = x$. Clearly, in additive groups, the role of the identity element is played by 0, and whenever we use "+" as the group operation, we will denote the identity element by "0". The axiom of the inverse then will say that the sum of an element and its inverse is zero. It is natural to denote inversion in an additive group as $-x$.

This gives us a wealth of examples of additive groups:

- integer numbers with addition operation $(\mathbb{Z}, +)$,

- rational numbers with addition operation $(\mathbb{Q}, +)$,

- real numbers with addition operation $(\mathbb{R}, +)$,

- any vector space with addition operation $(V, +)$.

In examples coming from the arithmetic, we also have the commutative law, $x \cdot y = y \cdot x$ (or $x + y = y + x$ for additive groups). The commutative law is not part of the axioms of a group, and many groups do not satisfy this law. Groups satisfying the commutative law are called *commutative* or *abelian*[1].

An example of a non-commutative group will be the set of invertible $N \times N$ matrices, which is called the general linear group and denoted $GL(N)$. The identity element in this group is the identity matrix I. This set is closed under multiplication since we have an explicit formula for the inverse of the product:

$$(AB)^{-1} = B^{-1}A^{-1}.$$

It may look strange that we changed the order of factors in the right hand side of the above formula. An easy way to understand the reason for this is to interpret it as the "Socks and Shoes Rule". In the morning, we first put on socks, then shoes. To undo this operation, we should take off the shoes first! There is also a way to verify this formula with algebra:

$$(AB)(B^{-1}A^{-1}) = A(BB^{-1})A^{-1} = AA^{-1} = I.$$

As we have seen earlier, product of matrices is non-commutative in general, hence $GL(N)$ is a non-commutative group.

We are going to discuss our next example in more detail. Consider the group G of symmetry transformations of a square. There are two

---

[1]In honour of the Norwegian mathematician Niels Abel

types of symmetry transformations – rotations and reflections. The identity transformation I may be considered to be a rotation in angle $0°$. Other rotations are counterclockwise rotations in angles $90°$, $180°$ and $270°$, which we will denote R_1, R_2 and R_3 respectively. The group G also contains four reflections, two reflections, T_1 and T_2 with axes through the midpoints of the opposite sides of the square, and two reflections, V_1 and V_2 with the axes through the pairs of opposite vertices.

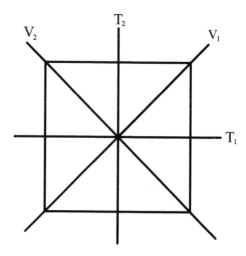

We assume that the first basis vector $\mathbf{e}_1$ of $\mathbb{R}^2$ is on the axis of T_1, while $\mathbf{e}_2$ is on the axis of T_2.

We can then compute the matrices of all 8 elements of G:

$$I = \begin{pmatrix} 1 & 0 \\ 0 & 1 \end{pmatrix}, \quad R_1 = \begin{pmatrix} 0 & -1 \\ 1 & 0 \end{pmatrix}, \quad R_2 = \begin{pmatrix} -1 & 0 \\ 0 & -1 \end{pmatrix}, \quad R_3 = \begin{pmatrix} 0 & 1 \\ -1 & 0 \end{pmatrix},$$

$$T_1 = \begin{pmatrix} 1 & 0 \\ 0 & -1 \end{pmatrix}, \quad T_2 = \begin{pmatrix} -1 & 0 \\ 0 & 1 \end{pmatrix}, \quad V_1 = \begin{pmatrix} 0 & 1 \\ 1 & 0 \end{pmatrix}, \quad V_2 = \begin{pmatrix} 0 & -1 \\ -1 & 0 \end{pmatrix}.$$

We wish to build a multiplication table for this group. It is easy to compute the products of rotations. For example, a $270°$ rotation, followed by a $180°$ rotation will yield a $90°$ rotation, hence $R_2 R_3 = R_1$, and so on. This will complete a quarter of the multiplication table:

	I	R_1	R_2	R_3	T_1	T_2	V_1	V_2
I	I	R_1	R_2	R_3				
R_1	R_1	R_2	R_3	I				
R_2	R_2	R_3	I	R_1				
R_3	R_3	I	R_1	R_2				
T_1								
T_2								
V_1								
V_2								

Next we point out that multiplication tables of groups possess a "Sudoku" property – no element can appear twice in the same row or column. This follows from the following algebraic argument: if an element appears twice in a row of element X, in columns Y and Z, we will have an equality $XY = XZ$. Multiplying both sides by X^{-1} on the left, we get $X^{-1}(XY) = X^{-1}(XZ)$. Applying the associative law, we get $(X^{-1}X)Y = (X^{-1}X)Z$. Using the axiom of the inverse, we obtain $eY = eZ$, and finally applying the axiom of the identity, we conclude $Y = Z$, which is a contradiction. The origin of this contradiction was an assumption that we have the same element appearing twice in the same row. Hence that cannot happen.

Since elements cannot repeat in the same row or column, each row/column must contain all elements of a group. This implies that the bottom left and upper right corners of the multiplication table of G must be filled with reflections, while the bottom right corner must contain rotations.

Hence, the product of two rotations is a rotation. The product of a rotation and a reflection, in either order, is a reflection. The product of two reflections is a rotation.

We can easily fill the first row and column, since these correspond to the identity element. We also know that a reflection applied twice yields the identity transformation. We also note that R_2, being a rotation by $180°$, transforms every vector into its opposite and has matrix $-I$. Since multiplication by $-I$ flips all signs in a matrix, regardless of the order of factors, the results are easy to calculate. We record this as follows:

	I	R_1	R_2	R_3	T_1	T_2	V_1	V_2
I	I	R_1	R_2	R_3	T_1	T_2	V_1	V_2
R_1	R_1	R_2	R_3	I				
R_2	R_2	R_3	I	R_1	T_2	T_1	V_2	V_1
R_3	R_3	I	R_1	R_2				
T_1	T_1		T_2		I			
T_2	T_2		T_1			I		
V_1	V_1		V_2				I	
V_2	V_2		V_1					I

Let us next calculate $T_1 R_1$ and $R_1 T_1$:

$$T_1 R_1 = \begin{pmatrix} 1 & 0 \\ 0 & -1 \end{pmatrix}\begin{pmatrix} 0 & -1 \\ 1 & 0 \end{pmatrix} = \begin{pmatrix} 0 & -1 \\ -1 & 0 \end{pmatrix} = V_2,$$

$$R_1 T_1 = \begin{pmatrix} 0 & -1 \\ 1 & 0 \end{pmatrix}\begin{pmatrix} 1 & 0 \\ 0 & -1 \end{pmatrix} = \begin{pmatrix} 0 & 1 \\ 1 & 0 \end{pmatrix} = V_1.$$

This shows that the group of symmetries of a square is non-commutative!
Placing these in the table,

	I	R_1	R_2	R_3	T_1	T_2	V_1	V_2
I	I	R_1	R_2	R_3	T_1	T_2	V_1	V_2
R_1	R_1	R_2	R_3	I	V_1			
R_2	R_2	R_3	I	R_1	T_2	T_1	V_2	V_1
R_3	R_3	I	R_1	R_2				
T_1	T_1	V_2	T_2		I			
T_2	T_2		T_1			I		
V_1	V_1		V_2				I	
V_2	V_2		V_1					I

we can use the "Sudoku" property of a multiplication table, to complete the spots for $T_1 R_3 = V_1$, $R_3 T_1 = V_2$, $T_2 R_1 = V_1$, $T_2 R_3 = V_2$,

$R_1 T_2 = V_2$, $R_3 T_2 = V_1$.

	I	R_1	R_2	R_3	T_1	T_2	V_1	V_2
I	I	R_1	R_2	R_3	T_1	T_2	V_1	V_2
R_1	R_1	R_2	R_3	I	V_1	V_2		
R_2	R_2	R_3	I	R_1	T_2	T_1	V_2	V_1
R_3	R_3	I	R_1	R_2	V_2	V_1		
T_1	T_1	V_2	T_2	V_1	I			
T_2	T_2	V_1	T_1	V_2		I		
V_1	V_1		V_2				I	
V_2	V_2		V_1					I

We can compute $V_1 R_1$ and $R_1 V_1$ using the associative law:

$$V_1 R_1 = (T_2 R_1) R_1 = T_2 (R_1 R_1) = T_2 R_2 = T_1,$$
$$R_1 V_1 = R_1 (R_1 T_1) = (R_1 R_1) T_1 = R_2 T_1 = T_2,$$

and complete the rest of these two corners using the "Sudoku" property:

	I	R_1	R_2	R_3	T_1	T_2	V_1	V_2
I	I	R_1	R_2	R_3	T_1	T_2	V_1	V_2
R_1	R_1	R_2	R_3	I	V_1	V_2	T_2	T_1
R_2	R_2	R_3	I	R_1	T_2	T_1	V_2	V_1
R_3	R_3	I	R_1	R_2	V_2	V_1	T_1	T_2
T_1	T_1	V_2	T_2	V_1	I			
T_2	T_2	V_1	T_1	V_2		I		
V_1	V_1	T_1	V_2	T_2			I	
V_2	V_2	T_2	V_1	T_1				I

We leave completion of the fourth corner of the table as an exercise.

Some groups that we may wish to study could be very large. For example, the number of elements in one important group, called the *Monster*, is

808,017,424,794,512,875,886,459,904,961,710,757,005,754,368,000,000,000.

If we decided to write down the multiplication table for the Monster group, we would run out of ink, since the number of entries in its multiplication table exceeds the number of atoms in the visible Universe!

We wish to develop more efficient algebraic methods to carry out computations in groups.

Let us generalize our previous example and consider the group D_n of symmetries of a regular polygon with n vertices. This group is called the *dihedral* group. It is easy to see that D_n has n rotations (in angles that are multiples of $360°/n$), and n reflections. Let us denote the counterclockwise rotation in angle $360°/n$ by R, and fix one of the reflections and call it T. Then all rotations in the dihedral group can be expressed as powers of R: R, R^2, R^3, ..., R^{n-1}, $R^n = e$. The same argument as we used in case of a square, shows that the product of a reflection with a rotation is a reflection. By the "Sudoku" principle, reflections T, TR, TR^2, ..., TR^{n-1} are distinct, and hence exhaust all reflections in the dihedral group D_n.

This allows us to list all the elements in the dihedral group as

$$D_n = \left\{ R^i, TR^i \,\middle|\, i = 0, 1, 2, \ldots, n-1 \right\}.$$

We note two relations: $R^n = e$ and $T^2 = e$. Since the product of R^i and T is a reflection, its square is the identity element, $(R^iT)^2 = e$, and then this element is its own inverse: $R^iT = (R^iT)^{-1} = T^{-1}R^{-i} = TR^{-i}$. It turns out that these three relations

$$R^n = e, \quad T^2 = e, \quad R^iT = TR^{-i},$$

allow us to multiply any two elements in the dihedral group. For example, let us compute in D_7 the product of TR^5 and TR^3:

$$(TR^5)(TR^3) = T(R^5T)R^3 = T(TR^{-5})R^3 = T^2R^{-2} = R^7R^{-2} = R^5.$$

9 Lagrange's Theorem

It could happen that inside a group there is another smaller group. For example, the set of rotations $\{R^i \mid i = 0, 1, \ldots, n-1\}$ inside the dihedral group D_n, is itself a group. This is an example of a *subgroup*.

Definition. Let G be a group. A subset H in G is called a *subgroup* if it satisfies the following three properties:
(1) the identity element e is in H,
(2) with every element h in H, its inverse, h^{-1} is also in H,
(3) with any two elements h_1, h_2 in H, the product $h_1 h_2$ is in H.

We can see that positive rational numbers $\mathbb{Q}_+$ form a subgroup in the multiplicative group $\mathbb{Q}^*$, and even integers form a subgroup in the additive group $\mathbb{Z}$. Another example of a subgroup would be $\{e, T\}$ in D_n. The one-element set $\{e\}$ and G itself are subgroups in G.

The set of orthogonal $N \times N$ matrices forms a subgroup, denoted $O(N)$, in the general linear group $GL(N)$.

Exercise. Determine all possible subgroups in D_4.

An important class of subgroups is *cyclic* subgroups.

Definition. Let g be an element in a group G. The cyclic subgroup generated by g is the set H of all integer powers of g:

$$H = \left\{ g^k \mid k = \ldots, -2, -1, 0, 1, 2, \ldots \right\}.$$

Clearly, this set is a subgroup since $g^k g^s = g^{k+s}$. If the group G is finite, the cyclic subgroup generated by element g cannot be infinite, and the list of powers of g will contain infinitely many repetitions. Let us analyze the structure of a cyclic subgroup in this case.

Definition. The order of an element g in a group G is the smallest positive integer number n such that $g^n = e$. If such a positive integer does not exist, we say that the order of g is infinity.

Examples. In group D_4 element T has order 2, and element R has order 4. In $\mathbb{Q}^*$ element $g = -1$ has order 2 since $(-1)^2 = 1$, while $g = 5$ has infinite order since no positive power of 5 equals 1.

Proposition. Let G be a finite group. Then any element g in G has a finite order. If n is the order of g then the cyclic subgroup generated by g has n elements $g^0 = e$, $g^1 = g$, g^2, g^3, ..., g^{n-1}.

Proof. To prove that g has a finite order, we need to show that there exists at least one positive integer k such that $g^k = e$. Since the set of all powers of g must contain repetitions, we will have $g^s = g^r$ for some integers $s < r$. Multiplying both sides by g^{-s}, we will get $g^{r-s} = e$ with $r - s$ being a positive integer. Hence g has a finite order.

Let n be the order of g. Then $g^n = e$, $g^{n+1} = g$, etc. Thus every positive power of g gets reduced to one of the elements $g^0, g^1, \ldots, g^{n-1}$. The list $\{g^0, g^1, \ldots, g^{n-1}\}$ can not have repetitions. Otherwise, we can apply the above argument and get $g^k = e$ with $0 < k < n$, which will contradict the definition of the order.

For the negative powers of g, we see that $g^{-1} = g^n g^{-1} = g^{n-1}$, $g^{-2} = g^{n-2}$, etc., and the claim of the Proposition follows.

Definition. A group G is called cyclic, if there exists an element g in G, such that integer powers of g exhaust G.

Definition. Let g be an element of a group G, and let H be a subgroup in G. A *coset* gH is a set of products $\{gh\}$ as h runs over the subgroup H.

Let us construct the cosets of the subgroup $H = \{e, R, R^2, R^3\}$ in the dihedral group D_4.

Clearly, the coset eH is just equal to H. The coset RH is also equal to H, since it contains the elements $\{R, R^2, R^3, R^4 = e\}$. The coset TH is $\{T, TR, TR^2, TR^3\}$, and actually for any i, the cosets $(TR^i)H$ and TH are the same. Here we can see that in this example there are just two distinct cosets, $\{e, R, R^2, R^3\}$ and $\{T, TR, TR^2, TR^3\}$.

It follows from the "Sudoku" property of multiplication that the size of any coset gH is equal to the number of elements in H.

Proposition. Let H be a subgroup in a group G. Then G is a union of non-overlapping cosets of H.

Proof. We need to show that distinct cosets do not overlap, that is, if two cosets aH and bH have a common element c then they are equal. Indeed, if c belongs to both cosets then $c = ah_1 = bh_2$ for some h_1, h_2 in H. Then $a = bh_2h_1^{-1}$. We need to show that any element of aH is in bH and vice versa. Let d belong to aH. Then for some h_3 in H, $d = ah_3 = bh_2h_1^{-1}h_3$. Since H is a subgroup, $h_2h_1^{-1}h_3$ in in H, and d is then in the coset bH. Likewise, every element of bH is in aH, so $aH = bH$.

We also see that the cosets cover the whole group since every element g belongs to its own coset gH.

Definition. The number of elements in a group G is called the *order* of a group G.

Lagrange's Theorem. Let G be a finite group.
(a) The order of any subgroup H in G is a divisor of the order of G.
(b) The order of any element g in G is a divisor of the order of G.

Proof. The group G is a union of non-overlapping cosets of H, which all have the same size. Hence

$$\text{Order of } G = \text{Order of } H \times \text{The number of cosets of } H.$$

This implies the claim of part (a) of the Theorem.

To prove part (b), we note that every element g generates a cyclic subgroup, whose order is equal to the order of the element g. Hence (b) follows from (a).

Corollary. Let G be a group of order n and let g be an element of G. Then $g^n = e$.

Proof. Let k be the order of g. By Lagrange's Theorem, $n = ks$ for some s. Then $g^n = (g^k)^s = e^s = e$.

10 Additive and Multiplicative Groups of Remainders

In this chapter we shall study addition and multiplication of remainders from the point of view of group theory.

The set $\mathbb{Z}_m$ of remainders after division by m forms a group with respect to addition. Its identity element is the zero remainder. In fact this group is cyclic with generator 1, since every remainder k in this group may be expressed as a sum of k ones: $1+1+\ldots+1$. The order of element 1 in $\mathbb{Z}_m$ is equal to m, since taking a sum of m ones will yield the identity element 0.

Consider as an example the group $\mathbb{Z}_{10}$. In this group element 2 has order 5, since $2+2+2+2+2 = 0 \bmod 10$, while the order of 3 is 10, as the smallest multiple of 3, which has remainder 0 mod 10 is $30 = 10 \times 3 = 3+3+3+3+3+3+3+3+3+3$. The following table lists the orders of elements in $\mathbb{Z}_{10}$:

x	0	1	2	3	4	5	6	7	8	9
order of x	1	10	5	10	5	2	5	10	5	10

Note that in agreement with Lagrange's Theorem, the order of every element is a divisor of 10, which is the order of the group $\mathbb{Z}_{10}$.

The answer for the general case is expressed in terms of least common multiples (LCM) and greatest common divisors (GCD).

Proposition. The order of element k in the additive group $\mathbb{Z}_m$ is

$$\frac{\mathrm{LCM}(m,k)}{k} = \frac{m}{\mathrm{GCD}(m,k)}.$$

Proof. Let r be the order of k in $\mathbb{Z}_m$. This means that rk is the smallest multiple of k, which is divisible by m. This implies that rk is the least common multiple of k and m, and we obtain the claim of the Proposition.

Next we consider the multiplicative structure of $\mathbb{Z}_m$. We can see that $\mathbb{Z}_m$ is *not* a group with respect to multiplication. To begin with, element 0 does not have a multiplicative inverse. It could also happen that some non-zero elements are not invertible either.

Definition. Remainders a and b in $\mathbb{Z}_m$ are called multiplicative inverses of each other if $ab = 1 \bmod m$.

For example, in $\mathbb{Z}_{10}$, elements 3 and 7 are multiplicative inverses of each other, since $3 \times 7 = 1 \bmod 10$, while 2 is not invertible because there is no remainder b for which $2b = 1 \bmod 10$.

Definition. The multiplicative group $\mathbb{Z}_m^*$ is the set of invertible remainders mod m.

Clearly, the set of invertible remainders forms a group, since the product of invertible remainders is invertible, $(ab)^{-1} = b^{-1}a^{-1}$.

We need to figure out which exactly remainders mod m are invertible. We are going to show below that a remainder k has a multiplicative inverse in $\mathbb{Z}_m$ if and only if $\text{GCD}(m,k) = 1$.

The proof of this fact, as well as the general method for computing the inverses, is based on the algorithm of computing the greatest common divisors, which goes back to Euclid. Let us now discuss the *Euclidean algorithm*.

One way to compute the greatest common divisor is to use prime factorization: for example, to compute $\text{GCD}(96,60)$, we may factor both numbers into powers of primes:

$$96 = 2^5 \times 3, \quad 60 = 2^2 \times 3 \times 5.$$

Then collecting common prime powers, we get $\text{GCD}(96, 60) = 2^2 \times 3 = 12$.

This method, however, becomes inefficient when the numbers in question have large prime factors. Let us try to decide, what is $\text{GCD}(4187, 2923)$? Factoring these numbers by hand will take some effort, and for larger numbers we do not know a good way of factoring even with a computer. In fact, the security of the RSA cryptosystem which we will discuss in a later chapter, is based precisely on the hardness of factorization of large integers.

Computing GCDs is an entirely different matter – we can efficiently compute these even for very large integers, thanks to Euclid.

The idea of the Euclidean algorithm is based on the fact that $\mathrm{GCD}(a, b) = \mathrm{GCD}(a - b, b)$. Indeed, if d is a common divisor of a, b, then it is also a common divisor of $a - b, b$, and vice versa. We can take this idea further by allowing to subtract from a arbitrary multiples of b, obtaining the following

Proposition. Divide a by b with a remainder: $a = sb + r$, where $0 \leq r < b$. Then $\mathrm{GCD}(a, b) = \mathrm{GCD}(b, r)$.

Applying this to the above example, we get that $\mathrm{GCD}(4187, 2923) = \mathrm{GCD}(2923, 1264)$. Repeating this process, we will get:

$$
\begin{array}{rcl|rcl}
4187 - 2923 & = & 1264 & \mathrm{GCD}(4187, 2923) & = & \mathrm{GCD}(2923, 1264) \\
2923 - 2 \times 1264 & = & 395 & \mathrm{GCD}(2923, 1264) & = & \mathrm{GCD}(1264, 395) \\
1264 - 3 \times 395 & = & 79 & \mathrm{GCD}(1264, 395) & = & \mathrm{GCD}(395, 79) \\
395 - 5 \times 79 & = & 0 & \mathrm{GCD}(395, 79) & = & 79.
\end{array}
$$

This tells us that $\mathrm{GCD}(4187, 2923) = 79$.

Reversing the calculations in the Euclidean algorithm, we obtain the following important result:

Theorem. Let $\mathrm{GCD}(a, b) = d$. Then there exist integers u, v such that

$$d = au + bv.$$

Applied to our example, this Theorem tells us that there exist integers u, v such that $4187u + 2923v = 79$ (clearly, one of the two numbers u, v must be negative). It is not at all obvious what are the values of u and v in this example. To get these, we run backwards the calculations we did in the Euclidean algorithm:

$$79 = 1264 - 3 \times 395$$
$$= 1264 - 3 \times (2923 - 2 \times 1264) = 7 \times 1264 - 3 \times 2923$$
$$= 7 \times (4187 - 2923) - 3 \times 2923 = 7 \times 4187 - 10 \times 2923,$$

concluding that $u = 7$, $v = -10$ is a desired solution.

Now we can prove the following

Theorem. A remainder k has a multiplicative inverse in $\mathbb{Z}_m$ if and only if $\mathrm{GCD}(m, k) = 1$.

Proof. Suppose $\mathrm{GCD}(m,k) = 1$. By the previous Theorem, there exist integers u, v, such that $1 = mu + kv$. Since $mu = 0 \mod m$, we get $1 = kv \mod m$, thus v is the multiplicative inverse of k in $\mathbb{Z}_m$.

To prove the converse, assume that k has a multiplicative inverse v in $\mathbb{Z}_m$, that is, $kv = 1 \mod m$. Two integers have the same remainder mod m whenever their difference is divisible by m. Thus $kv - 1 = ms$ for some s, and $kv - ms = 1$. If d is a common divisor of k and m, then it is also a divisor of $kv - ms$, which implies that d is a divisor of 1. Then the only common divisor of k and m is 1, which means that $\mathrm{GCD}(m,k) = 1$.

An important special case is when the modulus is a prime number p. In this case every non-zero remainder satisfies $\mathrm{GCD}(p,k) = 1$. The above Theorem then implies that in $\mathbb{Z}_p$ every non-zero remainder has a multiplicative inverse, so $\mathbb{Z}_p^*$ has $p - 1$ elements.

Fermat's Little Theorem. Let p be a prime number. If a is an integer not divisible by p then

$$a^{p-1} = 1 \mod p.$$

Proof. Apply Lagrange's Theorem to the multiplicative group $\mathbb{Z}_p^*$. Since the order of this group is $p - 1$, for every element a in $\mathbb{Z}_p^*$, we have $a^{p-1} = e$, which is exactly the claim of the Theorem.

Consider $p = 13$ as an example. Let us determine the order of $g = 2$ in $\mathbb{Z}_{13}^*$:

$$
\begin{array}{llll}
2^1 = 2 & 2^4 = 3 & 2^7 = 11 & 2^{10} = 10 \\
2^2 = 4 & 2^5 = 6 & 2^8 = 9 & 2^{11} = 7 \\
2^3 = 8 & 2^6 = 12 & 2^9 = 5 & 2^{12} = 1
\end{array}
$$

Hence the order of $g = 2$ in $\mathbb{Z}_{13}^*$ is 12, which also tells us that $\mathbb{Z}_{13}^*$ is a cyclic group of order 12 with $g = 2$ being a generator. We can calculate the orders of all elements in $\mathbb{Z}_{13}^*$. Since $3 = 2^4$ while $2^{12} = 1$, the order of 3 is 3.

Definition. An isomorphism of two groups is a one-to-one correspondence between their elements that preserves group operations.

The function $f(x) = 2^x$ is an isomorphism between the additive group $\mathbb{Z}_{12}$ and the multiplicative group $\mathbb{Z}_{13}^*$. It preserves group operations since $2^{x+y} = 2^x \times 2^y$.

We are going to state, without proof, the following result:

Theorem. Let p be a prime number. The group $\mathbb{Z}_p^*$ has a cyclic generator g, so that powers of g exhaust $\mathbb{Z}_p^*$. The function $f(x) = g^x$ is an isomorphism between $\mathbb{Z}_{p-1}$ and $\mathbb{Z}_p^*$.

There is no general rule that specifies which element of $\mathbb{Z}_p^*$ is a cyclic generator. Moreover, when g is given, the function $f(x) = g^x$ is a *trap-door function* (assuming that p is a large prime). This function is straightforward to compute, yet, it is difficult to compute the inverse of f, that is, given remainder h in $\mathbb{Z}_p^*$, determine the integer value of x, for which $h = g^x \mod p$. Since the function f is an exponential function, its inverse is called the *discrete logarithm function*, $x = \log_g(h)$. Of course, one could try to find x by exhaustive search, but for large primes that would be computationally infeasible.

There are several widely used cryptosystems which are based on the assumption that the discrete logarithm function is hard to compute. Shor's quantum algorithm not only can be used for integer factorization, but also has a version that computes discrete logarithms.

We conclude this chapter with

Chinese Remainder Theorem. Suppose $\mathrm{GCD}(m, s) = 1$. For any pair of remainders $a \mod m$, $b \mod s$, there exists a unique remainder $x \mod ms$, such that $x = a \mod m$ and $x = b \mod s$.

Example. A system

$$\begin{cases} x = 2 \mod 7 \\ x = 5 \mod 8 \end{cases}$$

has a unique solution in $\mathbb{Z}_{56}$, which is $x = 37$.

Proof. Since $\mathrm{GCD}(m, s) = 1$, there exist integers u, v, such that $mu + sv = 1$. Obviously,

$$mu \mod m = 0, \qquad sv \mod s = 0.$$

Reducing equality $mu + sv = 1 \mod m$ and $\mod s$, we get

$$sv \mod m = 1, \qquad mu \mod s = 1.$$

Finally, we construct the solution as $x = asv + bmu$. Let us verify that this formula produces the desired result:

$$x \mod m = a \times 1 + b \times 0 \mod m = a \mod m,$$
$$x \mod s = a \times 0 + b \times 1 \mod s = b \mod s.$$

Since the number of pairs of remainders $(a \mod m, b \mod s)$ is the same as the number of remainders in $\mathbb{Z}_{ms}$, a solution will be unique.

Corollary. Suppose $\mathrm{GCD}(m, s) = 1$. A remainder k is invertible in $\mathbb{Z}_{ms}$ if and only if $k \mod m$ is invertible in $\mathbb{Z}_m$ and $k \mod s$ is invertible in $\mathbb{Z}_s$.

Corollary. Suppose p and q are two primes with $p \neq q$. Then the order of the group $\mathbb{Z}_{pq}^*$ is $(p-1)(q-1)$.

Applying Lagrange's Theorem, we also get an analogue of Fermat's Little Theorem, which is used in RSA cryptosystem:

Theorem. Suppose p and q are two primes with $p \neq q$. Assume g is not divisible by p or q. Then

$$g^{(p-1)(q-1)} = 1 \mod pq.$$

11 Lie Groups

Lie groups (pronounced "lee") are named in honour of the Norwegian mathematician Sophus Lie who was the first to study them. Lie groups are symmetry groups that are continuous. A prototypical example of a Lie group is the group of rotations of a sphere. Fixing the axis of rotation, we can continuously change the angle of rotation. Alternatively, we may fix the angle of rotation, and continuously change the axis of rotation. Or, we can simultaneously change both the angle and the axis of rotation in a continuous way.

This contrasts the group of symmetries of a cube, where symmetries do not admit continuous deformations. If we try to rotate a cube by a small angle, it will not self-impose, so such rotations will not be symmetries of a cube.

Our goal is to understand groups $O(N)$ of orthogonal transformations. We begin with the case of orthogonal transformations of $\mathbb{R}^2$.

The group $O(2)$ is the group of symmetries of a circle. It has a presentation which is analogous to the one we used for the dihedral groups. The elements of $O(2)$ are of two types: rotations and reflections.

Recall the expression for the rotation matrices:

$$R_\alpha = \begin{pmatrix} \cos\alpha & -\sin\alpha \\ \sin\alpha & \cos\alpha \end{pmatrix}.$$

Let us also consider the reflection T with respect to the X-axis:

$$T = \begin{pmatrix} 1 & 0 \\ 0 & -1 \end{pmatrix}.$$

Then all other reflections may be constructed as products of T with all possible rotations:

$$T_\alpha = R_\alpha T = \begin{pmatrix} \cos\alpha & -\sin\alpha \\ \sin\alpha & \cos\alpha \end{pmatrix} \begin{pmatrix} 1 & 0 \\ 0 & -1 \end{pmatrix} = \begin{pmatrix} \cos\alpha & \sin\alpha \\ \sin\alpha & -\cos\alpha \end{pmatrix}.$$

Note that T_α is the reflection matrix we wrote down in chapter 6.

We get the following algebraic description of the group $O(2)$:

Theorem. The group $O(2)$ consists of the elements $\{R_\alpha, T_\alpha\}$, where α is the angle parameter, so that $R_{\alpha+360°} = R_\alpha$, $T_{\alpha+360°} = T_\alpha$. Multiplication in $O(2)$ is given as follows:

$$R_\alpha R_\beta = R_{\alpha+\beta}, \quad T_\alpha T_\beta = R_{\alpha-\beta}, \quad R_\alpha T_\beta = T_{\alpha+\beta}, \quad T_\alpha R_\beta = T_{\alpha-\beta}.$$

Proof. Since $T_{-\gamma}$ is a reflection, its inverse is itself: $R_{-\gamma}T = (R_{-\gamma}T)^{-1} = T^{-1}R_{-\gamma}^{-1} = TR_\gamma$. Using this relation we can compute:

$$T_\alpha R_\beta = R_\alpha T R_\beta = R_\alpha R_{-\beta} T = R_{\alpha-\beta} T = T_{\alpha-\beta}.$$

Verification of other relations is analogous and is left as an exercise.

We saw that the orthogonal group $O(2)$ consists of the elements of two types - rotations and reflections. What about orthogonal transformations in the 3-dimensional space? It turns out that elements of $O(3)$ are also of two types, one of which is spatial rotations (around some axis). The second type includes reflections with respect to a plane, but this is not all. In order to describe orthogonal transformations of a 3-dimensional space, we need one theorem from linear algebra. The proof of this theorem is not difficult, but involves the concept of eigenvalues. Since this is a more advanced topic in linear algebra, we omit the proof.

Theorem. Let T be an orthogonal transformation of $\mathbb{R}^3$. Then there exists a unit vector $\mathbf{v}$ such that either $T\mathbf{v} = \mathbf{v}$ or $T\mathbf{v} = -\mathbf{v}$.

Let $\mathbf{v}$ be a vector from the above Theorem, and let $\mathbf{u}$ be a vector orthogonal to $\mathbf{v}$. Since T is an orthogonal transformation, vectors $T\mathbf{v}$, $T\mathbf{u}$ are again orthogonal to each other. Taking into account that $T\mathbf{v} = \pm\mathbf{v}$, we conclude that $T\mathbf{u}$ is orthogonal to $\mathbf{v}$. As a result we see that the plane orthogonal to $\mathbf{v}$ is *invariant* under T, i.e., T transforms vectors from this plane into vectors on the same plane. Hence, restricted to the plane orthogonal to vector $\mathbf{v}$, T becomes a 2-dimensional orthogonal transformation, either a rotation of this plane, or a reflection with respect to a line in this plane.

In the argument below we use the fact that a reflection of a plane has a vector which is transformed to its opposite and a vector that is transformed into itself – the first vector is perpendicular to the line of reflection, while the second vector is on the line of reflection.

We consider 3 cases:

(1) T has a unit vector $\mathbf{v}$ such that $T\mathbf{v} = \mathbf{v}$, but does not have a unit vector $\mathbf{w}$ such that $T\mathbf{w} = -\mathbf{w}$.

(2) T has a unit vector $\mathbf{w}$ such that $T\mathbf{w} = -\mathbf{w}$, but does not have a unit vector $\mathbf{v}$ such that $T\mathbf{v} = \mathbf{v}$.

(3) T has both unit vectors $\mathbf{v}, \mathbf{w}$ such that $T\mathbf{v} = \mathbf{v}$, $T\mathbf{w} = -\mathbf{w}$.

In the first case, transformation T must be a rotation of a plane perpendicular to vector $\mathbf{v}$, since T does not transform any vector into its opposite. Hence T is a spatial rotation with $\mathbf{v}$ as its axis.

In the second case, T restricted to the plane perpendicular to $\mathbf{w}$ is again a rotation. Hence this transformation is a composition of a rotation of a plane together with a reflection in the same plane.

In case (3), we note that vectors $\mathbf{v}$ and $\mathbf{w}$ must be perpendicular to each other, since T preserves the dot product: $\mathbf{v} \cdot \mathbf{w} = T\mathbf{v} \cdot T\mathbf{w} = \mathbf{v} \cdot (-\mathbf{w}) = -\mathbf{v} \cdot \mathbf{w}$, which implies $\mathbf{v} \cdot \mathbf{w} = 0$. Take a unit vector $\mathbf{u}$, which is perpendicular to both $\mathbf{v}$ and $\mathbf{w}$. Then $T\mathbf{u}$ is perpendicular to $T\mathbf{v} = \mathbf{v}$ and $T\mathbf{w} = -\mathbf{w}$, which means that $T\mathbf{u}$ is proportional to $\mathbf{u}$. Since T preserves lengths, either $T\mathbf{u} = \mathbf{u}$, or $T\mathbf{u} = -\mathbf{u}$. We constructed an orthonormal basis $\{\mathbf{v}, \mathbf{u}, \mathbf{w}\}$ with T acting as follows:

$$(3a) \quad T\mathbf{v} = \mathbf{v}, \quad T\mathbf{u} = \mathbf{u}, \quad T\mathbf{w} = -\mathbf{w},$$

or

$$(3b) \quad T\mathbf{v} = \mathbf{v}, \quad T\mathbf{u} = -\mathbf{u}, \quad T\mathbf{w} = -\mathbf{w}.$$

In case (3a), T is a reflection with respect to a plane spanned by vectors $\mathbf{u}$ and $\mathbf{v}$. In case (3b), T is a rotation around $\mathbf{v}$ in angle $180°$.

Summarizing these results, we obtain:

Theorem. An orthogonal transformation of $\mathbb{R}^3$ is either a spatial rotation around some axis, or a composition of such a rotation with a reflection in the plane orthogonal to the axis of rotation.

A quantum algorithm is an orthogonal transformation of a 2^n-dimensional space of n-qubits. For this reason, we are interested in understanding the structure of orthogonal transformations of spaces of an arbitrary dimension. Their description is given by the following theorem, which we present without proof.

Before we state the theorem, as a warm-up, let us ask the following question: Is it possible to have two 2-dimensional planes in a 4-dimensional space which intersect at a single point? This may be challenging to visualize, since our intuition comes from a 3-dimensional space, where two intersecting planes must have a common line. Yet, in a 4-dimensional space, two planes may indeed intersect at a point. To illustrate this, let us call the coordinate axes $XYZW$. Then the XY plane consists of vectors with two last components being zero, while the vectors on the ZW plane have zero first two components. Clearly, the intersection of these 2-dimensional planes is just a single point – the origin.

Theorem. For any orthogonal transformation T of $\mathbb{R}^N$ one can find subspaces $L_1, L_2, \ldots, L_k$ in $\mathbb{R}^N$ such that:

(1) Each subspace L_i has dimension either 1 or 2, and their dimensions add up to N.

(2) All subspaces L_i are mutually orthogonal and intersect only at the origin,

(3) Each subspace L_i is *invariant* under T, that is, T transforms a vector from L_i into a vector in the same subspace L_i,

(4) If L_i is 1-dimensional, then $T\mathbf{v} = \mathbf{v}$ or $T\mathbf{v} = -\mathbf{v}$ for every vector $\mathbf{v}$ in L_i,

(5) If L_i is 2-dimensional, then T acts as a rotation transformation of the plane L_i.

Let us illustrate this Theorem with the case of $\mathbb{R}^4$. For a given orthogonal transformation T, we can use the subspaces $\{L_i\}$ to construct an orthonormal basis of $\mathbb{R}^4$, so that the matrix of T is of one of the following types:

(a) Diagonal matrix with ± 1 on the diagonal. This is the case when $\mathbb{R}^4$ decomposes into four 1-dimensional invariant subspaces:

$$T = \begin{pmatrix} \pm 1 & 0 & 0 & 0 \\ 0 & \pm 1 & 0 & 0 \\ 0 & 0 & \pm 1 & 0 \\ 0 & 0 & 0 & \pm 1 \end{pmatrix}.$$

(b) If $\mathbb{R}^4$ decomposes into a 2-dimensional invariant subspace and two 1-dimensional invariant subspaces, its matrix will look like:

$$T = \begin{pmatrix} \cos\alpha & -\sin\alpha & 0 & 0 \\ \sin\alpha & \cos\alpha & 0 & 0 \\ 0 & 0 & \pm1 & 0 \\ 0 & 0 & 0 & \pm1 \end{pmatrix}.$$

(c) And finally $\mathbb{R}^4$ may decompose into two 2-dimensional invariant subspaces. In this case T is a double rotation (try to visualize this!):

$$T = \begin{pmatrix} \cos\alpha & -\sin\alpha & 0 & 0 \\ \sin\alpha & \cos\alpha & 0 & 0 \\ 0 & 0 & \cos\beta & -\sin\beta \\ 0 & 0 & \sin\beta & \cos\beta \end{pmatrix}.$$

As we pointed out, a quantum algorithm is an orthogonal transformation of the 2^n-dimensional space of n-qubits. To implement it, we break it down into a sequence of more elementary transformations, each involving just one or two qubits. This is analogous to implementations of classical algorithms, which are also broken down into steps, each being some elementary operation. Algebraically, such a decomposition of a quantum algorithm is a factorization of a big orthogonal matrix in a product of certain elementary matrices.

The possibility of factorizations of this nature was first observed by Euler in 1774, who studied factorizations of orthogonal transformations in a 3-dimensional space. Euler proved the following

Theorem. Any rotation in a 3-dimensional space with coordinate axes XYZ may be factored as a composition of a rotation around the X axis in some angle α, followed by a rotation around the Y axis in some angle β, followed by a rotation around the X axis again in some angle γ.

The parameters α, β, γ are called the Euler angles of the given rotation. This Theorem tells us that it is enough to have the ability to make rotations just around the X and Y axes, in order to be able to generate an arbitrary rotation of $\mathbb{R}^3$.

Algebraically, this Theorem claims that any 3-dimensional rotation matrix may be factored in the following way:

$$\begin{pmatrix} 1 & 0 & 0 \\ 0 & \cos\gamma & -\sin\gamma \\ 0 & \sin\gamma & \cos\gamma \end{pmatrix} \begin{pmatrix} \cos\beta & 0 & -\sin\beta \\ 0 & 1 & 0 \\ \sin\beta & 0 & \cos\beta \end{pmatrix} \begin{pmatrix} 1 & 0 & 0 \\ 0 & \cos\alpha & -\sin\alpha \\ 0 & \sin\alpha & \cos\alpha \end{pmatrix}.$$

To prove this Theorem, it is much easier to give a geometric argument, rather than algebraic. Fix a rotation T. Suppose T transforms some orthogonal axes $X_0 Y_0 Z_0$ into the standard coordinate axes XYZ. We point out that it is sufficient to only keep track of two coordinate axes. If we have a rotation that transforms X_0 into X and Y_0 into Y, then automatically Z_0 will be transformed in Z, since rotations preserve the angles between the axes.

The goal of the first step is to rotate $X_0 Y_0 Z_0$ around the X axis, into $X_1 Y_1 Z_1$ making the new X_1 axis perpendicular to the Y axis. This is possible since we can rotate any non-zero vector around the X axis to make the result be in the XZ plane.

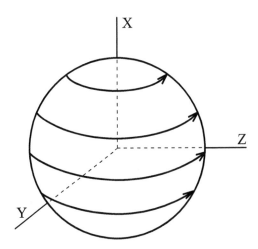

The second step is to perform a rotation around the Y axis, transforming $X_1 Y_1 Z_1$ into $X_2 Y_2 Z_2$ with the new axis X_2 aligned with the X axis. This is possible since we got X_1 in the XZ plane on the previous step. Since the axis Y_2 is perpendicular to X, it is in the YZ plane.

Finally, we perform a rotation around the X axis, to align the new axis Y_3 with Y. Since now X_3 is aligned with X, and Y_3 is aligned with Y, then automatically Z_3 is aligned with Z.

12 RSA Cryptosystem

In chapter 3, we discussed cryptosystems based on a shared secret key. For modern Internet commerce these secret key cryptosystems are not practical, since the two parties of a transaction did not have prior contact. Still, the use of encryption is essential for Internet commerce, because the global network is organized in such a way that communications pass through a chain of servers which are owned by various third party entities, and the exact path of a message cannot even be predicted in advance.

The solution to this challenge is to use *public key cryptography*. Public key cryptosystems allow anyone to encrypt messages, but only the intended recipient is able to decrypt them. Imagine that Bob, an owner of an online store, wishes to receive credit card numbers from his customers in a secure way. Bob will generate two keys: a *public key*, that is required for encryption, and a *secret key*, which is used for decryption. Public key is made known to everyone, it is usually embedded into the web page of Bob's store. Then Alice, a customer at Bob's store, will use the public key to encrypt her credit number and will send the encrypted message to Bob, in order to make a payment for her purchase. Usually the encryption is done by Alice's browser, which will know how to read Bob's public key from his web page, and is programmed to run the encryption algorithm.

Bob will receive an encrypted transmission from Alice, and will decrypt Alice's credit card number using his secret key. Since only Bob has the secret key, he is the only person capable of decrypting Alice's message.

One of the first public key cryptosystems was proposed by Rivest, Shamir and Adleman in 1974. This cryptosystem, which is known as RSA, is based on factoring of integers. Given two large prime numbers, p and q, it is easy to compute their product $N = pq$. However, if we are given N and are told that it is a product of two large primes, finding the factors is hard.

The public key for RSA cryptosystem is a pair of integers, $N = pq$ and k, where k is an integer with a property $\text{GCD}(k, (p-1)(q-1)) = 1$.

For example, we may choose k to be a prime number in the interval $\max(p-1, q-1) < k < (p-1)(q-1)$.

The secret key is a set of three integers, p, q and s, where s is computed as

$$s = k^{-1} \mod (p-1)(q-1).$$

As we are going to see below, any digital data may be converted into a sequence of remainders mod N: $m_1, m_2, \ldots$. Alice will encrypt each *plaintext* remainder m into a *cyphertext* remainder c using the following encryption formula:

$$c = m^k \mod N,$$

and will send the cyphertext sequence $c_1, c_2, \ldots$ to Bob.

In order to decrypt a cyphertext block c, Bob will compute $c^s \mod N$. Let us verify that in this way Bob will recover the plaintext block m.

Indeed, since $ks = 1 \mod (p-1)(q-1)$, we can write $ks = 1 + r(p-1)(q-1)$ for some integer r. Then computing mod N, we get:

$$c^s = (m^k)^s = m^{ks} = m^{1+r(p-1)(q-1)}$$

$$= m^1 \times (m^{(p-1)(q-1)})^r = m \times 1^r = m \mod N.$$

Here we used the analogue of Fermat's Little Theorem for remainders mod $N = pq$:

$$m^{(p-1)(q-1)} = 1 \mod pq.$$

One should point out that the last formula only holds when $\mathrm{GCD}(m, pq) = 1$. In practice, for large primes p, q, the probability of this equality failing is so tiny, that it can be ignored for all practical purposes. In fact, finding a single non-zero remainder m with $\mathrm{GCD}(m, pq) \neq 1$ will yield factoring of pq, since the value of this GCD will be either p or q. Since we rely on the fact that factoring is not possible in a reasonable amount of time, we can just as well assume that accidental failure of our encryption scheme will not occur.

If someone can factor N into p and q, then the product $(p-1)(q-1)$ can be computed and the secret decryption key may be found:

$$s = k^{-1} \mod (p-1)(q-1).$$

Proposition. Let N be a product of two unknown primes, $N = pq$. Let $\mathbb{Z}_N^*$ be the multiplicative group of invertible remainders mod N. Knowing value of the order $M = (p-1)(q-1)$ of this group would allow us to effectively factor $N = pq$.

Proof. First note that $M = (p-1)(q-1) = N - p - q + 1$, so if we know the values of N and M we also know the value of $S = p + q = N - M + 1$.

Let us construct a quadratic equation that has p and q as roots:

$$(X - p)(X - q) = X^2 - (p+q)X + pq = X^2 - SX + N.$$

Even though we do not know p and q, we know the coefficients S and N of this quadratic polynomial. Applying the standard formula for the roots of the quadratic equation

$$X^2 - SX + N = 0,$$

we find the values of the factors p, q.

Let us discuss implementation details of the RSA cryptosystem. First of all, how to convert arbitrary data into a sequence of remainders? Any digital data is stored on a computer as a file, which is a long string of 0's and 1's. All we have to do is to figure out how to convert a long binary string into a sequence of remainders mod N.

Choose n in such a way that $2^n < N - 1 \leq 2^{n+1}$. We take our long binary string and cut it into blocks of length n. Each block can viewed as a binary form of an integer between 0 and $2^n - 1$. We add 2 to this integer, so that the result is between 2 and $2^n + 1$ and may be interpreted as a remainder mod N. The reason for adding 2 is to avoid remainders 0 and 1.

Another technical issue worth investigating is the complexity of computing $c^s \bmod N$. Here the magnitude of s is comparable to N, so it is very large. Let us ask a question: how many multiplications are required to compute $c^{2018} \bmod N$? At a first glance it looks like we need 2017 multiplications to compute this, however it turns out that we can do much better. Let us begin by writing down a binary expansion of 2018:

$$2018 = 1024+512+256+128+64+32+2 = 2^{10}+2^9+2^8+2^7+2^6+2^5+2^1.$$

It takes one multiplication to square a number. By repeated squaring we may compute c^2, $(c^2)^2 = c^4$, $(c^4)^2 = c^{2^3}$, ..., $(c^{2^9})^2 = c^{2^{10}}$. This takes 10 multiplications. Then we use the binary expansion of 2018 to express:

$$c^{2018} = c^{2^{10}} \times c^{2^9} \times c^{2^8} \times c^{2^7} \times c^{2^6} \times c^{2^5} \times c^2.$$

Thus we only need a total of 16 multiplications to compute c^{2018}. In general, we need less than $2 \log_2 s$ multiplications to compute c^s.

One more issue to pay attention to is the size of the numbers involved in these calculations. For large s, the number c^s is huge. Fortunately, we only need to compute $c^s \bmod N$. If we take a remainder mod N after each multiplication, we will never deal with integers exceeding N^2.

13 Parallel Classical Computations with a Quantum Computer

Any classical computation may be viewed as a function taking n-bit strings as input and k-bit strings as output:

$$f : \ B_n \to B_k.$$

Here we denote the set of all binary strings of length n as B_n.

As a simple example, we may consider a function that computes the square of a binary 3-bit integer, $f : \ B_3 \to B_6$, $f(101) = 011001$ ($5^2 = 25$). Here we need to take 6-bit integers as output since f applied to 111 produces a 6-bit result 110001.

This interpretation is valid not only for the functions of arithmetic nature, but is applicable to all computer programs. After all, any digital data may be converted to a binary string format (this is how files are stored on a computer, from text files to movies). Any computer program has digital data as input and output, and may be thus viewed as a function on binary strings.

We would like to understand, how can we code a classical computation with a quantum computer. An obstacle we face is that quantum algorithms are supposed to be invertible whereas classical computations are not. To overcome this, we convert a classical function $f : \ B_n \to B_k$ into an invertible classical function

$$\widehat{f} : \ B_{n+k} \to B_{n+k}.$$

The idea here is to combine together input and output bits, and set

$$\widehat{f}(x, y) = (x, y \oplus f(x)).$$

Here x is the n-bit input, y is the initial value for the output bits, and we denote by the symbol $\oplus$ bitwise addition mod 2 (without carry).

We merge together the bits allocated for the input and the output of function f and make this joint set of bits to be both the input and the output space for the new function $\widehat{f}$. The convention here is that the values of the bits corresponding to the old input do not change.

In applications, the values of the bits corresponding to the old output are initialized to a zero string. Then we can use $\widehat{f}$ to compute f:

$$\widehat{f}(x,0) = (x, f(x)).$$

For example if $f : B_3 \to B_6$ is a function that computes the square of a 3-bit integer, then

$$\widehat{f}(101, 000000) = (101, 011001).$$

Here put put a comma between x and y just for convenience of reading, these two parts form a single 9-bit string. Even though we are mostly interested in computations where y is initialized to a zero string, $\widehat{f}$ is defined with all 9-bit arguments, for example,

$$\widehat{f}(101, 111000) = (101, 100001).$$

To compute the value of the last 6 bits, we first evaluated $f(101) = 011001$, and then performed bitwise addition mod 2 with the initial state of the output bits:

$$111000 \oplus 011001 = 100001.$$

Proposition. Let f be an arbitrary classical computation $f : B_n \to B_k$. Then the function $\widehat{f} : B_{n+k} \to B_{n+k}$, defined as

$$\widehat{f}(x, y) = (x, y \oplus f(x)),$$

is invertible. The inverse of $\widehat{f}$ is itself.

Proof. We need to verify that $\widehat{f}(\widehat{f}(x, y)) = (x, y)$. Indeed,

$$\widehat{f}(\widehat{f}(x, y)) = \widehat{f}(x, y \oplus f(x)) = (x, y \oplus f(x) \oplus f(x)) = (x, y).$$

On the last step we used the fact that any binary string added to itself using bitwise addition mod 2, produces a zero string.

Recall that a quantum computation is an orthogonal linear transformation of the space of n-qubits. Thus we need to convert an invertible classical computation into a linear transformation. To define a linear transformation, we need to specify its values on the basis vectors, and the basis of the space of n-qubits is given by the binary strings of length n.

To a classical function $f : B_n \to B_k$ we associate a linear transformation T_f of the space of $(n + k)$-qubits defined by

$$T_f(|z\rangle) = |\widehat{f}(z)\rangle.$$

Note that T_f is a very special kind of a linear transformation – it transforms every basis vector into another basis vector. Typically, a linear transformation will transform a basis vector into a linear combination of basis vectors.

Transformation T_f is invertible since $\widehat{f}$ is. Moreover, $T_f^{-1} = T_f$.

Since T_f permutes vectors in a basis, the images of the basis vectors are orthogonal to each other and all have unit length. Hence, T_f is an orthogonal linear transformation.

For example, for the function $f : B_3 \to B_6$ we considered above, transformation T_f will be a linear transformation of a 512-dimensional space of 9-qubits, and can be represented by a 512×512 matrix (for obvious reasons we are not going to write it down). Since T_f permutes basis vectors, its matrix will have a single entry 1 in each row and column.

The definition of T_f shows the possibility of implementation of classical computations with a quantum computer. Naturally, a quantum computer is capable of doing more than that. After all, we only use permutation matrices to implement classical computations, and we have many more matrices available to us.

Still, there is one new feature that we gain by doing classical computations with a quantum computer. Imagine that we use a state

$$\frac{1}{\sqrt{2^n}} \sum_{x=0}^{2^n-1} |x\rangle |0\rangle$$

as input to T_f. Here we write x as an n-bit binary integer, and 0 represents a string of k zeros. The factor $\frac{1}{\sqrt{2^n}}$ in front is to make this vector have norm 1.

We get

$$T_f \left(\frac{1}{\sqrt{2^n}} \sum_{x=0}^{2^n-1} |x\rangle \, |0\rangle \right) = \frac{1}{\sqrt{2^n}} \sum_{x=0}^{2^n-1} |x\rangle \, |f(x)\rangle \, .$$

Now the new state incorporates information about the values of $f(x)$ for all possible x. To achieve something like that with a classical computer, we would need to compute values of $f(x)$ sequentially, and we would need to perform 2^n iterations. This will be prohibitively long even for modest values of n (after all, the age of the Universe is less than 2^{70} milliseconds).

A quantum computer will compute this in a single iteration. This happens since we organized the input as a superposition of all possible classical inputs. Then, a quantum computation, being a linear transformation, will generate as output a superposition of states that will contain information about all possible values of f. This shows that quantum computations are massively parallel.

The drawback here is that the information about the values of f cannot be easily extracted from the output state. Imagine that we wish to solve an equation $f(x) = 0$, and assume for simplicity that there is one value $x = a$ for which this equation holds. Then we would know that one of the terms in the output will be $|a\rangle \, |0\rangle$, but we cannot get access to the value of a by a measurement. If we perform a measurement on the output state, we will only observe the value $|b\rangle \, |f(b)\rangle$ for a random input b. For more details on this proplem, see the description of Grover's quantum algorithm in the Appendix.

The challenge in designing quantum algorithms is to organize computations in such a way, that a measurement will give us the answer to our problem with a high enough probability.

14 Nuts and Bolts of Classical Computations

When a computer program is executed, it is broken down into a sequence of basic operations that are executed by the computer processor. Today's processors are powerful and can execute fairly complex operations. For our analysis, we want to go to the roots of computation, and discuss the most basic operations that are sufficient for building a computer. These basic operations are: AND, OR and NOT. The first two operations take two bits as arguments, whereas NOT is an operation on a single bit.

The origins of these operations are in logic. Bit values are interpreted as True/False with $x = 1$ interpreted as "x is True" and $x = 0$ interpreted as "x is False". Then x AND y is interpreted as "both x and y are True", while x OR y is interpreted as "either x is True, or y is True, or both". These are called operations of the Boolean logic, in honour of George Boole, a 19th century mathematician who laid down the foundations of symbolic logic.

We have the following value tables for these logic operations:

x	y	x AND y
0	0	0
0	1	0
1	0	0
1	1	1

x	y	x OR y
0	0	0
0	1	1
1	0	1
1	1	1

x	NOT x
0	1
1	0

We are going to use shorter notations, $x \wedge y$ for x AND y, $x \vee y$ for x OR y, $\overline{x}$ for NOT x. Function AND is also called *conjunction*, while function OR is called *disjunction*.

Proposition. Operations of Boolean logic satisfy the following properties:

(1) $\overline{x \wedge y} = \overline{x} \vee \overline{y}$,

(2) $\overline{x \vee y} = \overline{x} \wedge \overline{y}$,

(3) $\overline{\overline{x}} = x$,

(4) $x \wedge (y \vee z) = (x \wedge y) \vee (x \wedge z)$,

(5) $x \vee (y \wedge z) = (x \vee y) \wedge (x \vee z)$.

We leave the proof of this Proposition as an exercise.

When it comes to the order of operations, it is customary to carry out AND before OR, just as multiplication has a higher priority compared to addition. Thus, expression $x \wedge y \vee z \wedge w$ is understood as $(x \wedge y) \vee (z \wedge w)$, whereas $x \vee y \wedge z \vee w$ is understood as $x \vee (y \wedge z) \vee w$.

Example. Let us express the function $x = y$ (i.e., x is True if and only if y is True). This means that either both x and y are True, or both x and y are False. This can be expressed as

$$x \wedge y \vee \overline{x} \wedge \overline{y}.$$

Example. Let us express the following function: 2-bit binary integer $x_1 x_0$ is greater than 2-bit binary integer $y_1 y_0$. This function should produce value 1 if $x_1 x_0 > y_1 y_0$, and 0 otherwise.

We see that $x_1 x_0 > y_1 y_0$ if the first bits satisfy $x_1 > y_1$, or if $x_1 = y_1$ and $x_0 > y_0$. Inequality $x_1 > y_1$ means that $x_1 = 1$ and $y_1 = 0$, whereas the expression for the equality was found in the previous example. Combining these, we get an expression for this function in the language of Boolean algebra:

$$x_1 \wedge \overline{y}_1 \vee (x_1 \wedge y_1 \vee \overline{x}_1 \wedge \overline{y}_1) \wedge x_0 \wedge \overline{y}_0.$$

As we pointed out above, any classical computation may be viewed as a function $f : B_n \to B_k$. Such a function f may be replaced with k functions $f_1, f_2, \ldots, f_k$, where each f_i is defined on B_n and produces values in $B_1 = \{0, 1\}$. Here f_1 computes the first bit of the value of f, f_2 the second bit, and so on. For this reason we are focusing our attention on functions

$$f : B_n \to B_1.$$

We claim that elementary Boolean functions {AND, OR, NOT} allow us to express arbitrary Boolean functions. In electronics we implement Boolean functions with electrical circuits, where presence of voltage in a wire represents 1, and no voltage represents 0. Imagine that we have a supply of boxes that implement elementary functions {AND, OR, NOT} (the box for AND will have two wires going in, and a single wire going out). The consequence of the previous claim is that using these elementary boxes (also called *gates*) we can build a computer processor. In fact, this is exactly how we build computers today, only instead of using separate boxes for each elementary Boolean function, we assemble them with millions of these elementary gates on a single chip.

Let us show that we can use elementary Boolean functions {AND, OR, NOT} to express the function $f : B_3 \to B_1$, given by the following value table:

x	y	z	f
0	0	0	0
0	0	1	1
0	1	0	0
0	1	1	1
1	0	0	1
1	0	1	0
1	1	0	0
1	1	1	0

Consider the second row of the value table, which says that f assumes value 1 when $x = 0, y = 0, z = 1$. If we consider a conjunction $\overline{x} \wedge \overline{y} \wedge z$, it also generates value value 1 when $x = 0, y = 0, z = 1$, but assumes value 0 for all other values of x, y, z.

Look at other rows where function f assumes values 1. This happens when $x = 0, y = 1, z = 1$ and when $x = 1, y = 0, z = 0$. The conjunctions that generate value 1 for these values of x, y, z are $\overline{x} \wedge y \wedge z$, $x \wedge \overline{y} \wedge \overline{z}$. If we then take a disjunction

$$(\overline{x} \wedge \overline{y} \wedge z) \vee (\overline{x} \wedge y \wedge z) \vee (x \wedge \overline{y} \wedge \overline{z}),$$

it will generate exactly the same values as given in the value table of f. This expression is called the *disjunctive normal form* of f.

It is clear that this approach works for any Boolean function. We obtain the following

Theorem. Any Boolean function $f : B_n \to B_1$ may be expressed using the elementary Boolean functions {AND, OR, NOT} as a disjunctive normal form.

When we design a logical circuit of a computer, we should try to minimize the number of logical gates that we use, in order to reduce cost, energy consumption and to increase the speed of computations. Disjunctive normal forms, while being capable of implementing any Boolean function, are not usually optimal from the point of view of the number of gates they use. We illustrate this with the following example.

Consider a function f that implements addition of two 2-bit integers: $f(x_1x_0, y_1y_0) = x_1x_0 + y_1y_0 = z_2z_1z_0$. This function takes 4 bits as input (we separate the first two from the last two bits with a comma simply for convenience of reading), and produces 3 bits of output. We need to use 3 bits for the output because a sum of two 2-bit integers may be a 3-bit integer, for example: 3+2=5, which is written in binary as $11 + 10 = 101$, which means that $f(1110) = 101$.

x_1	x_0	y_1	y_0	z_2	z_1	z_0
0	0	0	0	0	0	0
0	0	0	1	0	0	1
0	0	1	0	0	1	0
0	0	1	1	0	1	1
0	1	0	0	0	0	1
0	1	0	1	0	1	0
0	1	1	0	0	1	1
0	1	1	1	1	0	0
1	0	0	0	0	1	0
1	0	0	1	0	1	1
1	0	1	0	1	0	0
1	0	1	1	1	0	1
1	1	0	0	0	1	1
1	1	0	1	1	0	0
1	1	1	0	1	0	1
1	1	1	1	1	1	0

Using addition of integers in binary form, we can easily complete the value table for f. We can view function f as three functions computing values of individual bits, $z_2(x_1x_0, y_1y_0)$, $z_1(x_1x_0, y_1y_0)$, $z_0(x_1x_0, y_1y_0)$. Let us determine complexity of their disjunctive normal forms. As a measure of complexity let us choose the number of conjunctions and disjunctions in the given expression. For simplicity, we are going to neglect negations (they are less expensive to implement in hardware). From the above table we see that value 1 is assumed in 8 rows for the functions z_1 and z_0, and 6 times for the function z_2. Hence the disjunctive normal forms for z_1 and z_0 have complexity $8 \times 3 + 7 = 31$, and the disjunctive normal form for z_2 has complexity $6 \times 3 + 5 = 23$.

The problem with the disjunctive normal forms is that they do not take into account any internal logic that may be present in a given function. For example, z_0 computes the last bit of a sum, which indicates its parity. However the parity of a sum is determined by the parities of the arguments: the sum of two even numbers is even, the sum of an even number and an odd number is odd, the sum of two odd numbers is even. Hence, z_0 is determined by values of x_0 and y_0 only, and we can give the following expression for it:

$$z_0 = (\overline{x}_0 \wedge y_0) \vee (x_0 \wedge \overline{y}_0).$$

The complexity of the latter expression is 3, which is much better than 31.

When we need to add two large numbers, we do not use "addition tables", instead we use the method of long addition. The same method works for integers in binary form, for example, the following computation

		1	1	1			1	
		1	0	1	1	0	1	
+		1	1	1	0	0	1	
	1	1	0	0	1	1	0	

represents addition 45+57=102 computed in binary. The carries are indicated in a small font above the top row.

We represent the above Boolean formula for addition of two bits mod 2 with the following circuit:

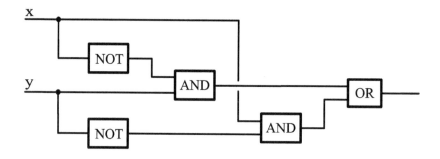

To simplify the diagrams, we will draw this circuit as a single addition box. Then the Boolean function $a + b + c \mod 2$ may be realized as a circuit in the following way:

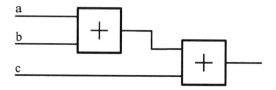

Another operation that we need in order to implement long addition is computation of a carry that appears when we add 3 bits (two input bits and a carry from the previous position). Note that in addition $a + b + c$ the carry occurs when at least two of the bits have value 1, so the carry function may be expressed as $(a \wedge b) \vee (a \wedge c) \vee (b \wedge c)$ with complexity 5, and may be further simplified to $a \wedge (b \vee c) \vee (b \wedge c)$ with complexity 4. The circuit for the carry computation is:

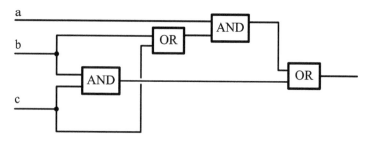

Now we can present the circuit for the addition of 2-bit binary integers. To simplify the diagram, we will be using the circuits for $a + b + c \bmod 2$, and for the carry as singles boxes

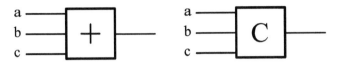

Note that the box for the computation of a carry in $a + b$, is just $a \wedge b$.

Finally, we get the circuit for the computation of z_2, z_1, z_0 in $x_1 x_0 + y_1 y_0 = z_2 z_1 z_0$.

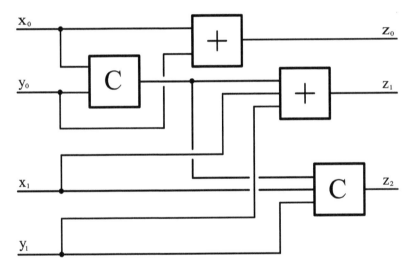

Evaluating complexity of each gate, we see that the total complexity of this circuit is $1 + 3 + 6 + 4 = 14$, which is much better than $31 + 31 + 23 = 85$, given by disjunctive normal forms.

15 Quantum Gates and Circuits

In this chapter we are going to show how to transform a classical logical circuit into a quantum circuit that implements classical computations on a quantum computer.

As we discussed earlier, quantum computations must be invertible, whereas classical gates {AND, OR} are not invertible. We fix this problem by combining input and output bits together and creating reversible analogues {RAND, ROR}, each being a function taking 3 bits of input and producing 3 bits of output:

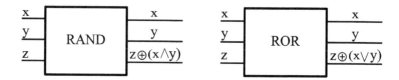

This creates gates that compute $z = x \wedge y$ and $z = x \vee y$ respectively, provided that the output bit z is initialized to 0. These gates are invertible and double application of a gate will restore the values of all variables.

The NOT gate is already invertible and compatible with quantum computations, hence there is no need to modify it.

There is one more operation that we need to add to our toolbox. In classical computations we can make copies of data. Copying/duplication operation is not invertible. In order to create the quantum analogue of copying, we need to apply the method of combining input and output to the function $f(x) = x$. In this way we obtain the gate which is called CNOT:

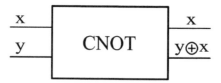

This gate performs copying of a classical bit x, provided that the output bit y is initialized to zero: $\text{CNOT}(x, 0) = (x, x)$. The name CNOT stands for "Controlled NOT" – this is how this gate is used in electronics, if $x = 1$ then the value of y is negated, and if $x = 0$ the value of y is unchanged.

Let us point out that CNOT only copies classical bit values, but not arbitrary quantum states. Let us see what happens when we apply CNOT to a superposition of classical states:

$$\text{CNOT}(a\,|0\rangle + b\,|1\rangle)\,|0\rangle = a\text{CNOT}(|00\rangle) + b\text{CNOT}(|10\rangle) = a\,|00\rangle + b\,|11\rangle .$$

Copying a quantum state would mean the following transformation:

$$(a\,|0\rangle + b\,|1\rangle)\,|0\rangle \mapsto (a\,|0\rangle + b\,|1\rangle)(a\,|0\rangle + b\,|1\rangle),$$

which differs from the action of CNOT. In fact there is a theorem claiming that copying unknown quantum states is impossible.

Notice also that in the above example, CNOT creates entanglement. The original state $(a\,|0\rangle + b\,|1\rangle)\,|0\rangle$ is factored in a product, and hence is not entangled, whereas the state $a\,|00\rangle + b\,|11\rangle$ is entangled, provided that $a, b \neq 0$.

Now it becomes straightforward to turn any classical circuit into a quantum circuit. Let us show how this is done on the example of the function $f(x, y) = x + y \bmod 2$. We transform the classical circuit given in the previous chapter into a quantum circuit using RAND, ROR, NOT, CNOT gates:

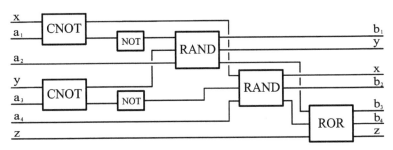

It is true that for the function $f(x, y) = x + y \bmod 2$ we could have given a much simpler quantum circuit using CNOT gates only.

The point of this exercise was to show how any classical computation circuit can be translated into a quantum circuit.

Note that in the quantum circuit we use not just the input and output variables, but also auxiliary variables for storing the results of intermediate calculations. This is quite similar to the way we do classical computations - we use extra internal memory to carry out the calculations.

When we develop software, we use compilers to convert code written in a high-level computer language into a code executable by the processor. Complex software is organized in modules, and once a module completes its task, a compiler releases auxiliary memory used by this module so that it becomes available to other modules. This task is known as *garbage collection*.

It is desirable to perform garbage collection with quantum computations as well, for two reasons. First we may want to reuse auxiliary qubits when they are no longer needed, and secondly, our output bits may get entangled with the auxiliary bits, and anything happening to the auxiliary bits may affect the values of the output qubits.

Quantum garbage collection would mean that auxiliary qubits which are initialized to zero, will have zero values at the end of the computation.

Theorem. For any classical computation there exists a quantum circuit which implements this computation with quantum garbage collection.

Consider a quantum implementation of a classical Boolean function $f : B_n \rightarrow B_k$. The qubit space used for this computation is divided into input qubits, output qubits and auxiliary qubits.

The quantum circuit T_f will transform $|x\rangle_{in} |0\rangle_{aux} |0\rangle_{out}$ into $|x\rangle_{in} |j(x)\rangle_{aux} |f(x)\rangle_{out}$. Here $|\ \rangle_{in}$ represents input qubits, $|\ \rangle_{aux}$ auxiliary qubits, $|\ \rangle_{out}$ output qubits, and $j(x)$ is the junk value of the auxiliary qubits produced by the quantum circuit.

To implement quantum garbage collection, we double the size of the output space, creating new qubits $|\ \rangle_{new\ out}$ with the same number of qubits as in $|\ \rangle_{out}$, and we now view the old output space $|\ \rangle_{out}$ as part of the auxiliary qubits. Now we run the following 3-step quantum computation:

Step 1. Apply T_f, transforming

$$|x\rangle_{in} |0\rangle_{aux} |0\rangle_{out} |0\rangle_{new\ out} \quad \text{into} \quad |x\rangle_{in} |j(x)\rangle_{aux} |f(x)\rangle_{out} |0\rangle_{new\ out} \cdot$$

Step 2. Use CNOT to copy from $|\ \rangle_{out}$ to $|\ \rangle_{new\ out}$, obtaining

$$|x\rangle_{in} |j(x)\rangle_{aux} |f(x)\rangle_{out} |f(x)\rangle_{new\ out} \cdot$$

Step 3. Apply T_f^{-1}, obtaining

$$|x\rangle_{in} |0\rangle_{aux} |0\rangle_{out} |f(x)\rangle_{new\ out} \cdot$$

We see that all auxiliary bits, including old output bits, are reset to 0.

In order to implement T_f^{-1}, we build a quantum circuit, where all quantum gates of T_f are placed in reverse order. Since each elementary quantum gate is its own inverse, this will undo T_f.

Here we modify the quantum circuit for $x + y$ mod 2, implementing quantum garbage collection:

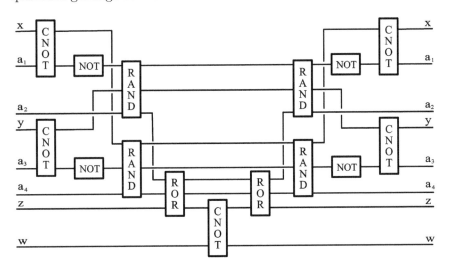

16   Discrete Fourier Transform

Fourier transform is a mathematical tool for studying periodic or nearly periodic functions. A good example of a nearly periodic function is the sound wave produced by a musical instrument. Let us look at the simplest musical device – a vibrating string. A plucked string vibrates, moving air around it.

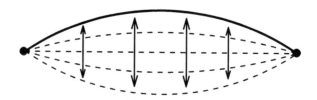

This creates a sound wave, periodically spaced pockets of air of higher density, interlaced with pockets of air of lower density. These pockets propagate in space away from the source of the sound. To record sound, we use a microphone, which has a flexible membrane inside. Once the sound wave hits the microphone, pockets of dense air push membrane to cave in, and pockets of low pressure will pull membrane out.

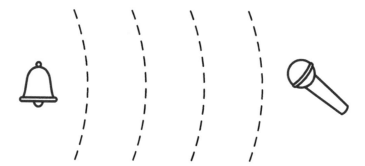

A microphone will transform vibration of a membrane into an electrical signal, which follows the motion of the membrane. In analog sound recording, this electrical signal may be used to magnetize a tape, so that the intensity of the magnetic field on the tape will follow the profile of the sound signal. In digital sound recording, electric signal produced by the microphone is *sampled*, capturing values of this signal at regular time intervals and storing them on a computer. For example, CD sound recordings are sampled at a rate of 44,100 Hertz, which means that one second of sound signal is stored as 44,100 values, usually scaled to be in the interval between -1 and 1.

Let us return to the discussion of a vibrating string. A string vibrating in a way shown on the picture above, will produce sound at a certain frequency, which is called its *base frequency* ω. It turns out, however, that the same string may vibrate in a more complicated way. It also has *higher vibrational modes*, which produce frequencies 2ω, 3ω, 4ω, etc. In practice, vibration of a string incorporates multiple higher vibrational modes in addition to the base mode.

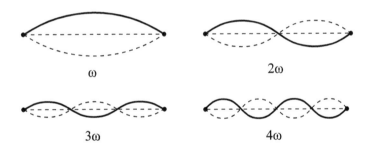

In music, higher vibrational modes present in the sound of an instrument, are called overtones. It is the distribution of intensities of overtones that distinguishes one musical instrument from another, playing the same note.

Here we present a plot of a sound of a flute, showing approximately 0.01 seconds of the recording. Notice how the profile is nearly periodic, with 5 periods that can be identified on this plot.

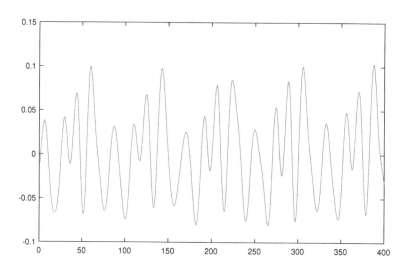

The next plot shows the frequency spectrum of this sound wave.

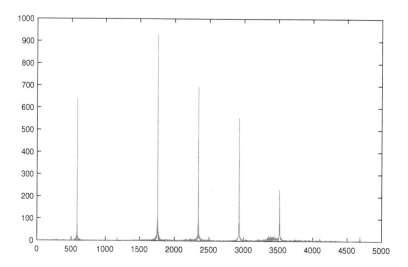

Because this is the sound of a musical instrument, playing a single note, the frequency plot exhibits spikes at the base frequency (587 Hertz, the note is D5), and at the multiples of the base frequency. Note that for this instrument, the second overtone is virtually absent from the frequency spectrum.

Analysis of the frequency spectrum of the signal, presented above, is carried out using discrete Fourier transform (DFT), which we are going to discuss next.

Let us consider a function $f(t)$ defined on an interval $0 \leq t \leq \pi$. We want to sample $f(t)$ at N equally spaced points $t_0, t_1, t_2, \ldots, t_{N-1}$ with the distance $\frac{\pi}{N}$ between the points. However, the first point t_0 is placed not at 0, but with an offset, $t_0 = \frac{\pi}{2N}$. Then $t_1 = t_0 + \frac{\pi}{N} = \frac{3\pi}{2N}$, $t_2 = t_1 + \frac{\pi}{N} = \frac{5\pi}{2N}$, etc., with the general formula

$$t_j = \frac{(2j+1)\pi}{2N}.$$

The last point will be close to the right end of the interval: $t_{N-1} = \frac{(2N-1)\pi}{2N} = \pi - \frac{\pi}{2N}$. By sampling $f(t)$ at these points we get an N-component vector $\mathbf{f} = (f_0, f_1, \ldots, f_{N-1})$, where $f_j = f(t_j)$. We are going to assume that N is even, $N = 2M$.

Waves are modelled with periodic functions $\cos(t)$ and $\sin(t)$. From now on we take the arguments of trigonometric functions to be in radians. Let us consider a family of functions

$$u_0(t) = \cos(t), \ u_1(t) = \cos(3t), \ldots, u_{M-1}(t) = \cos((N-1)t),$$

$$v_0(t) = \sin(t), \ v_1(t) = \sin(3t), \ldots, v_{M-1}(t) = \sin((N-1)t).$$

We can write these in a general form as

$$u_p(t) = \cos((2p+1)t), \ v_p(t) = \sin((2p+1)t), \quad p = 0, 1, \ldots, M-1.$$

We then use sampling to pass from these continuous functions to their discrete versions, creating an N-component vector from each of these functions:

$$\widetilde{\mathbf{u}}_0 = (\cos(t_0), \cos(t_1), \cos(t_2), \ldots, \cos(t_{N-1})),$$

$$\widetilde{\mathbf{u}}_1 = (\cos(3t_0), \cos(3t_1), \cos(3t_2), \ldots, \cos(3t_{N-1})),$$

$$\widetilde{\mathbf{u}}_2 = (\cos(5t_0), \cos(5t_1), \cos(5t_2), \ldots, \cos(5t_{N-1})),$$

$$\ldots$$

and analogously for the functions $v_p(t)$.

Substituting the values for the points t_j, we get

$$\widetilde{\mathbf{u}}_p =$$
$$\left(\cos\left(\frac{(2p+1)\pi}{2N}\right), \cos\left(\frac{(2p+1)3\pi}{2N}\right), \ldots, \cos\left(\frac{(2p+1)(2N-1)\pi}{2N}\right)\right),$$

$$\widetilde{\mathbf{v}}_p =$$
$$\left(\sin\left(\frac{(2p+1)\pi}{2N}\right), \sin\left(\frac{(2p+1)3\pi}{2N}\right), \ldots, \sin\left(\frac{(2p+1)(2N-1)\pi}{2N}\right)\right),$$

with $p = 0, 1, \ldots, M-1$.

This gives us a family of N vectors in $\mathbb{R}^N$. Let us study their properties.

Theorem. Vectors $\{\widetilde{\mathbf{u}}_0, \widetilde{\mathbf{u}}_1, \ldots, \widetilde{\mathbf{u}}_{M-1}, \widetilde{\mathbf{v}}_0, \widetilde{\mathbf{v}}_1, \ldots, \widetilde{\mathbf{v}}_{M-1}\}$ are orthogonal to each other.

Before we prove this Theorem, let us review trigonometric identities.

When we look at the properties of the trigonometric functions, it is important to keep in mind that $\cos\alpha$ is the X-coordinate of the point on a unit circle, corresponding to angle α, while $\sin\alpha$ is the Y-coordinate of the same point.

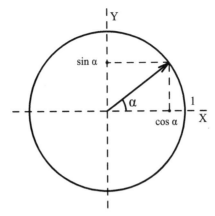

From this definition we see immediately the following properties:

$$\cos(-\alpha) = \cos\alpha, \quad \sin(-\alpha) = -\sin\alpha.$$

The only two trigonometric formulas that really should be memorized are:

$$\cos(\alpha + \beta) = \cos\alpha\cos\beta - \sin\alpha\sin\beta,$$

$$\sin(\alpha + \beta) = \sin\alpha\cos\beta + \cos\alpha\sin\beta.$$

All other identities may be then derived. Switching the sign of β in the above formulas, we get

$$\cos(\alpha - \beta) = \cos\alpha\cos\beta + \sin\alpha\sin\beta,$$

$$\sin(\alpha - \beta) = \sin\alpha\cos\beta - \cos\alpha\sin\beta.$$

Combining these formulas, we obtain

$$\cos(\alpha + \beta) + \cos(\alpha - \beta) = 2\cos\alpha\cos\beta,$$

$$\cos(\alpha - \beta) - \cos(\alpha + \beta) = 2\sin\alpha\sin\beta,$$

$$\sin(\alpha + \beta) + \sin(\alpha - \beta) = 2\sin\alpha\cos\beta.$$

We will also need the following

Proposition. Assume $\sin\alpha \neq 0$. Then

$$\cos(\alpha) + \cos(3\alpha) + \cos(5\alpha) + \ldots + \cos((2N-1)\alpha) = \frac{\sin(2N\alpha)}{2\sin\alpha},$$

$$\sin(\alpha) + \sin(3\alpha) + \sin(5\alpha) + \ldots + \sin((2N-1)\alpha) = \frac{1 - \cos(2N\alpha)}{2\sin\alpha}.$$

To prove the first identity of the Proposition, let us multiply its the left hand side by $2\sin\alpha$ and apply the formula for $2\sin\alpha\cos\beta$:

$$2\sin\alpha\cos(\alpha) + 2\sin\alpha\cos(3\alpha) + 2\sin\alpha\cos(5\alpha)$$
$$+ \ldots + 2\sin\alpha\cos((2N-1)\alpha)$$
$$= (\sin(2\alpha) - \sin(0)) + (\sin(4\alpha) - \sin(2\alpha)) + (\sin(6\alpha) - \sin(4\alpha))$$
$$+ \ldots + (\sin(2N\alpha) - \sin((2N-2)\alpha)).$$

Here most terms will cancel out, leaving $\sin(2N\alpha)$. Dividing both sides by $2\sin\alpha$, we get the first identity.

We leave the proof of the second identity of the Proposition as an exercise.

To prove the Theorem, we need to compute the dot products between the vectors in our family.

$$\tilde{\mathbf{u}}_p \cdot \tilde{\mathbf{u}}_s = \sum_{j=0}^{N-1} \cos\left(\frac{(2p+1)(2j+1)\pi}{2N}\right) \cos\left(\frac{(2s+1)(2j+1)\pi}{2N}\right).$$

Applying the formula for the product of cosines, we get

$$\frac{1}{2}\sum_{j=0}^{N-1} \cos\left(\frac{(2p+2s+2)(2j+1)\pi}{2N}\right) + \cos\left(\frac{(2p-2s)(2j+1)\pi}{2N}\right).$$

Next, using the previous Proposition, we evaluate the sums:

$$\frac{1}{2}\sum_{j=0}^{N-1} \cos\left(\frac{(p+s+1)(2j+1)\pi}{N}\right) = \frac{\sin\left(\frac{(p+s+1)2N\pi}{N}\right)}{4\sin\left(\frac{(p+s+1)\pi}{N}\right)}.$$

This computation is only valid if the sine in the denominator is non-zero. Indeed, since $0 \leq p, s \leq M - 1$, we conclude that $0 < \frac{(p+s+1)\pi}{N} < \pi$ and the sine value is non-zero. Evaluating the second sum we get

$$\frac{1}{2} \sum_{j=0}^{N-1} \cos\left(\frac{(p-s)(2j+1)\pi}{N}\right) = \frac{\sin\left(\frac{(p-s)2N\pi}{N}\right)}{4\sin\left(\frac{(p-s)\pi}{N}\right)},$$

provided that the sine in the denominator is non-zero. Here the denominator will turn into zero only when $p = s$. Note that both numerators turn into zero, hence $\tilde{\mathbf{u}}_p \cdot \tilde{\mathbf{u}}_s = 0$ when $p \neq s$. If $p = s$, the first sum is still zero, and all cosines in the second sum have value 1, giving $\tilde{\mathbf{u}}_p \cdot \tilde{\mathbf{u}}_p = \frac{N}{2}$.

Let us now compute the dot product $\tilde{\mathbf{u}}_p \cdot \tilde{\mathbf{v}}_s$:

$$\tilde{\mathbf{u}}_p \cdot \tilde{\mathbf{v}}_s = \sum_{j=0}^{N-1} \cos\left(\frac{(2p+1)(2j+1)\pi}{2N}\right) \sin\left(\frac{(2s+1)(2j+1)\pi}{2N}\right).$$

Applying the formula for $\sin \alpha \cos \beta$, we get

$$\frac{1}{2} \sum_{j=0}^{N-1} \sin\left(\frac{(2p+2s+2)(2j+1)\pi}{2N}\right) + \sin\left(\frac{(2s-2p)(2j+1)\pi}{2N}\right).$$

Using the second identity from the Proposition above, we simplify the sums:

$$\frac{1}{2} \sum_{j=0}^{N-1} \sin\left(\frac{(p+s+1)(2j+1)\pi}{N}\right) = \frac{1 - \cos\left(\frac{(p+s+1)2N\pi}{N}\right)}{4\sin\left(\frac{(p+s+1)\pi}{N}\right)},$$

with the denominator being non-zero. Since the numerator turns into zero, this sum vanishes. Evaluating the second sum we get

$$\frac{1}{2} \sum_{j=0}^{N-1} \sin\left(\frac{(s-p)(2j+1)\pi}{N}\right) = \frac{1 - \cos\left(\frac{(s-p)2N\pi}{N}\right)}{4\sin\left(\frac{(s-p)\pi}{N}\right)},$$

provided that $p \neq s$, in which case the sum is zero. If $p = s$, this sum is still zero since each summand is zero.

We conclude that $\tilde{\mathbf{u}}_p \cdot \tilde{\mathbf{v}}_s = 0$ for all p, s. We leave the calculation of $\tilde{\mathbf{v}}_p \cdot \tilde{\mathbf{v}}_s$ as an exercise.

Summarizing:

$$\tilde{\mathbf{u}}_p \cdot \tilde{\mathbf{u}}_s = \tilde{\mathbf{v}}_p \cdot \tilde{\mathbf{v}}_s = \begin{cases} N/2, & \text{if } p = s, \\ 0, & \text{if } p \neq s. \end{cases} \qquad \tilde{\mathbf{u}}_p \cdot \tilde{\mathbf{v}}_s = 0 \text{ for all } p, s.$$

This completes the proof of the Theorem.

To turn this set of vectors into an orthonormal basis, we divide each vector by its length:

$$\mathbf{u}_p = \sqrt{\frac{2}{N}}\tilde{\mathbf{u}}_p, \qquad \mathbf{v}_p = \sqrt{\frac{2}{N}}\tilde{\mathbf{v}}_p.$$

We can expand any vector $\mathbf{f} = (f_0, f_1, \ldots, f_{N-1})$ in this basis:

$$\mathbf{f} = \sum_{p=0}^{M-1} a_p \mathbf{u}_p + b_p \mathbf{v}_p.$$

The coefficients a_p and b_p in this expression are called the *Fourier coefficients* of $\mathbf{f}$. Discrete Fourier transform is a transformation of a vector of samples into a vector of Fourier coefficients:

$$(f_0, f_1, \ldots, f_{N-1}) \mapsto (a_0, a_1, \ldots, a_{M-1}, b_0, b_1, \ldots, b_{M-1}).$$

The original vector of samples describes the evolution of the signal in time. Each Fourier coefficient corresponds to a particular frequency. We say that Fourier transform is a transformation from *time domain* to *frequency domain*.

We need to solve the problem of expanding a given vector $\mathbf{f}$ as a linear combination of $\{\mathbf{u}_0, \mathbf{u}_1, \ldots, \mathbf{u}_{M-1}, \mathbf{v}_0, \mathbf{v}_1, \ldots, \mathbf{v}_{M-1}\}$. It turns out that it is much easier to do this for bases that are orthonormal.

Proposition. Let $\{\mathbf{w}_1, \mathbf{w}_2, \ldots, \mathbf{w}_N\}$ be an orthonormal basis of $\mathbb{R}^N$ and let $\mathbf{f}$ be a vector in $\mathbb{R}^N$. The coefficients of the expansion of $\mathbf{f}$ into a linear combination

$$\mathbf{f} = c_1 \mathbf{w}_1 + c_2 \mathbf{w}_2 + \ldots + c_N \mathbf{w}_N$$

may be found using dot products:

$$c_j = \mathbf{f} \cdot \mathbf{w}_j \quad \text{for} \quad j = 1, 2, \ldots, N.$$

Proof. Take the dot product of both sides of the linear combination with $\mathbf{w}_j$:

$$\mathbf{f} \cdot \mathbf{w}_j = c_1 \mathbf{w}_1 \cdot \mathbf{w}_j + c_2 \mathbf{w}_2 \cdot \mathbf{w}_j + \ldots + c_N \mathbf{w}_N \cdot \mathbf{w}_j.$$

Since the basis is orthonormal, all dot products in the right hand side will turn into zero, except $\mathbf{w}_j \cdot \mathbf{w}_j$, which is equal to 1. This will give us $\mathbf{f} \cdot \mathbf{w}_j = c_j$, as claimed.

This Proposition immediately yields the formulas for the Fourier coefficients:

$$a_p = \mathbf{f} \cdot \mathbf{u}_p = \sqrt{\frac{2}{N}} \sum_{j=0}^{N-1} f_j \cos\left(\frac{(2p+1)(2j+1)\pi}{2N}\right),$$

$$b_p = \mathbf{f} \cdot \mathbf{v}_p = \sqrt{\frac{2}{N}} \sum_{j=0}^{N-1} f_j \sin\left(\frac{(2p+1)(2j+1)\pi}{2N}\right),$$

with $p = 0, 1, \ldots, M - 1$.

If we know the Fourier coefficients $(a_0, a_1, \ldots, a_{M-1}, b_0, b_1, \ldots, b_{M-1})$, we can reconstruct the original signal $\mathbf{f} = (f_0, f_1, \ldots, f_{N-1})$. This procedure is called the *inverse Fourier transform*.

Since

$$\mathbf{f} = \sum_{p=0}^{M-1} a_p \mathbf{u}_p + b_p \mathbf{v}_p,$$

we can compute f_j by taking j-th component of each vector:

$$f_j = \sqrt{\frac{2}{N}} \sum_{p=0}^{M-1} a_p \cos\left(\frac{(2p+1)(2j+1)\pi}{2N}\right) + b_p \sin\left(\frac{(2p+1)(2j+1)\pi}{2N}\right).$$

Both direct and inverse discrete Fourier transforms are linear transformations of $\mathbb{R}^N$. Under the inverse DFT, vector $(1, 0, 0, \ldots, 0)$, which corresponds to $a_0 = 1$, gets transformed into the vector $\mathbf{u}_0$. Similarly, the images of the standard basis vectors $\mathbf{e}_k$ are the vectors $\{\mathbf{u}_0, \ldots, \mathbf{u}_{M-1}, \mathbf{v}_0, \ldots, \mathbf{v}_{N-1}\}$. Since these vectors are orthonormal, we conclude that inverse DFT is an orthogonal linear transformation. An inverse of an orthogonal linear transformation is orthogonal. This implies that direct DFT is also an orthogonal linear transformation.

This is good news for us, because it means that discrete Fourier transform is compatible with the paradigm of quantum computing. A quantum version of DFT, which is called *quantum Fourier transform* (QFT), is an essential ingredient of Shor's algorithm. We shall introduce QFT in a later chapter.

The magnitude of Fourier coefficients indicate how strongly a certain frequency is present in a given signal. In particular, when we apply the Fourier transform to a periodic signal, we shall see spikes in Fourier coefficients a_p and b_p at values of p that correspond to the multiples of the base frequency of the signal. This is exactly what we saw in the DFT plot for the flute recording shown above.

Let us consider a particularly simple example of a periodic signal, which is important for the Shor's algorithm. Fix two integers $0 \le s < m$, and consider the following sequence of length N and period m:

$$f_i = \begin{cases} 1, & \text{if } i = s \bmod m, \\ 0, & \text{otherwise.} \end{cases}$$

Let us assume that N/m is fairly large. We see that $f_i = 1$ for $i = s + jm$, with $j = 0, 1, \ldots, L - 1$. Here L is the smallest integer that is greater or equal to $(N - s)/m$.

Exercise. Compute the discrete Fourier transform for this sequence $(f_0, f_1, \ldots, f_{N-1})$. Show that Fourier coefficients are given by the following formulas:

$$a_p = \sqrt{\frac{2}{N}} \, \frac{\sin\left(\frac{(2p+1)(2s+(2L-1)m+1)\pi}{2N}\right) - \sin\left(\frac{(2p+1)(2s-m+1)\pi}{2N}\right)}{2\sin\left(\frac{(2p+1)m\pi}{2N}\right)},$$

$$b_p = \sqrt{\frac{2}{N}}\ \frac{\cos\left(\frac{(2p+1)(2s-m+1)\pi}{2N}\right) - \cos\left(\frac{(2p+1)(2s+(2L-1)m+1)\pi}{2N}\right)}{2\sin\left(\frac{(2p+1)m\pi}{2N}\right)}.$$

Hint: The first Proposition of this chapter may come handy in this computation.

Let us analyze when the Fourier coefficients obtained in this exercise have large values. Absolute values of the numerators in these formulas cannot exceed 2, since absolute values of sine and cosine do not exceed 1. Thus the only possibility for these Fourier coefficients to become large is when the denominator $\sin\left(\frac{(2p+1)m\pi}{2N}\right)$ has a value close to zero. This happens when $\frac{(2p+1)m}{2N}$ is close to an integer:

$$\frac{(2p+1)m}{2N} \approx K,$$

equivalently,

$$p \approx K \times \frac{N}{m} - \frac{1}{2}.$$

We see that the spikes in the values of the Fourier coefficients are located at the values of p that are multiples of the base frequency

$$\omega = \frac{N}{m}$$

(neglecting a small shift by $\frac{1}{2}$). Note that locations of the spikes are determined by the period m, and do not dependent on s.

Discrete Fourier transform is widely used in digital signal processing, and in particular for compression of audio files (MP3) and images (JPEG). Let us outline the idea of audio compression.

A typical sound signal is oscillating and the vast majority of its samples are not close to zero. If we perform a discrete Fourier transform, a majority of Fourier coefficients will be very close to zero, since a typical sound wave is localized to several fairly narrow frequency bands. We can replace small Fourier coefficients with zeros and then store/transmit only significant Fourier coefficients, reducing the size of data. The inverse DFT is used to reconstruct the signal from the stored Fourier coefficients. This will introduce small distortions to the recording, yet significant compression factors may be achieved with only a minor loss in quality.

17 Fast Fourier Transform

Fast Fourier transform (FFT) is a clever fast algorithm for computing discrete Fourier transform. Our goal is the same as in the previous chapter, to compute the Fourier coefficients:

$$a_p = \mathbf{f} \cdot \mathbf{u}_p = \sqrt{\frac{2}{N}} \sum_{j=0}^{N-1} f_j \cos\left(\frac{(2p+1)(2j+1)\pi}{2N}\right),$$

$$b_p = \mathbf{f} \cdot \mathbf{v}_p = \sqrt{\frac{2}{N}} \sum_{j=0}^{N-1} f_j \sin\left(\frac{(2p+1)(2j+1)\pi}{2N}\right).$$

Let us estimate the complexity of the computation, given by these formulas. When defining the computational complexity here, we will count the number of multiplications and ignore additions, since additions are far less expensive. We assume that all sine and cosine values are precomputed and stored. We will also ignore the factor of $\sqrt{\frac{2}{N}}$ since we can choose $\frac{N}{2}$ to be a power of 4, and resulting division by a power of 2 is a very simple operation when performed in binary.

Computation of each Fourier coefficient using the above formulas will then involve N multiplications, and there are N Fourier coefficients to be computed. Hence the complexity of DFT when computed with these formulas is N^2. Fast Fourier transform lets us compute exactly same thing, but with complexity $2N\log_2(N)$. For a vector of length 1024 (which is not uncommon in digital signal processing), the straightforward method will have complexity of $1,048,576$ versus $20,480$ for the FFT.

Fast Fourier transform is a *recursive* algorithm. Its idea is to split the given vector $\mathbf{f}$ into two vectors of half length each, apply FFT to each shorter vector and then recombine two sequences of Fourier coefficients for the two halves into a sequence of Fourier coefficients for $\mathbf{f}$.

From now on, we shall assume that N is a power of 2, $N = 2^n$, $M = N/2 = 2^{n-1}$.

Let us splice vector $\mathbf{f} = (f_0, f_1, f_2, \ldots, f_{N-1})$ into two shorter vectors as follows: $\mathbf{f} = (g_0, h_0, g_1, h_1, \ldots, g_{M-1}, h_{M-1})$. Here no computations are needed, we are only introducing new notations:

$$g_i = f_{2i}, \quad h_i = f_{2i+1}, \quad i = 0, 1, \ldots, M - 1.$$

This produces two vectors of length 2^{n-1}:

$$\mathbf{g} = (g_0, g_1, \ldots, g_{2^{n-1}-1}), \quad \mathbf{h} = (h_0, h_1, \ldots, h_{2^{n-1}-1}).$$

Let us apply Fourier transform to vectors $\mathbf{g}$ and $\mathbf{h}$:

$$\mathbf{g} \mapsto (a_0^g, \ldots, a_{2^{n-2}-1}^g, b_0^g, \ldots, b_{2^{n-2}-1}^g),$$

$$\mathbf{h} \mapsto (a_0^h, \ldots, a_{2^{n-2}-1}^h, b_0^h, \ldots, b_{2^{n-2}-1}^h).$$

Fourier coefficients for $\mathbf{g}$ and $\mathbf{h}$ are given by

$$a_s^g = \frac{1}{\sqrt{2^{n-2}}} \sum_{i=0}^{2^{n-1}-1} g_i \cos\left(\frac{(2s+1)(2i+1)\pi}{2^n}\right),$$

$$b_s^g = \frac{1}{\sqrt{2^{n-2}}} \sum_{i=0}^{2^{n-1}-1} g_i \sin\left(\frac{(2s+1)(2i+1)\pi}{2^n}\right),$$

with analogous formulas for a_s^h and b_s^h.

Now let us see how we can construct Fourier coefficients a_p, b_p for vector $\mathbf{f}$ from a_s^g, a_s^h, b_s^g and b_s^h.

$$a_p = \frac{1}{\sqrt{2^{n-1}}} \sum_{j=0}^{2^n-1} f_j \cos\left(\frac{(2p+1)(2j+1)\pi}{2^{n+1}}\right)$$

$$= \frac{1}{\sqrt{2^{n-1}}} \sum_{\substack{i=0 \\ j=2i}}^{2^{n-1}-1} f_{2i} \cos\left(\frac{(2p+1)(4i+1)\pi}{2^{n+1}}\right)$$

$$+ \frac{1}{\sqrt{2^{n-1}}} \sum_{\substack{i=0 \\ j=2i+1}}^{2^{n-1}-1} f_{2i+1} \cos\left(\frac{(2p+1)(4i+3)\pi}{2^{n+1}}\right)$$

$$= \frac{1}{\sqrt{2^{n-1}}} \sum_{i=0}^{2^{n-1}-1} g_i \cos\left(\frac{(2p+1)(4i+2)\pi}{2^{n+1}} - \frac{(2p+1)\pi}{2^{n+1}}\right)$$

$$+ \frac{1}{\sqrt{2^{n-1}}} \sum_{i=0}^{2^{n-1}-1} h_i \cos\left(\frac{(2p+1)(4i+2)\pi}{2^{n+1}} + \frac{(2p+1)\pi}{2^{n+1}}\right)$$

$$= \frac{1}{\sqrt{2^{n-1}}} \sum_{i=0}^{2^{n-1}-1} g_i \cos\left(\frac{(2p+1)(2i+1)\pi}{2^n}\right) \cos\left(\frac{(2p+1)\pi}{2^{n+1}}\right)$$

$$+ \frac{1}{\sqrt{2^{n-1}}} \sum_{i=0}^{2^{n-1}-1} g_i \sin\left(\frac{(2p+1)(2i+1)\pi}{2^n}\right) \sin\left(\frac{(2p+1)\pi}{2^{n+1}}\right)$$

$$+ \frac{1}{\sqrt{2^{n-1}}} \sum_{i=0}^{2^{n-1}-1} h_i \cos\left(\frac{(2p+1)(2i+1)\pi}{2^n}\right) \cos\left(\frac{(2p+1)\pi}{2^{n+1}}\right)$$

$$- \frac{1}{\sqrt{2^{n-1}}} \sum_{i=0}^{2^{n-1}-1} h_i \sin\left(\frac{(2p+1)(2i+1)\pi}{2^n}\right) \sin\left(\frac{(2p+1)\pi}{2^{n+1}}\right)$$

$$= \frac{a_p^g + a_p^h}{\sqrt{2}} \cos\left(\frac{(2p+1)\pi}{2^{n+1}}\right) + \frac{b_p^g - b_p^h}{\sqrt{2}} \sin\left(\frac{(2p+1)\pi}{2^{n+1}}\right).$$

We leave the computation for b_p as an exercise:

$$b_p = \frac{b_p^g + b_p^h}{\sqrt{2}} \cos\left(\frac{(2p+1)\pi}{2^{n+1}}\right) + \frac{-a_p^g + a_p^h}{\sqrt{2}} \sin\left(\frac{(2p+1)\pi}{2^{n+1}}\right).$$

Keep in mind that in fast Fourier transform algorithm, Fourier coefficients for subsequences **g** and **h** are again computed recursively, by splicing each of these into shorter subsequences. The original Fourier transform formulas are only used for the sequences of length 2, $(f_0, f_1) \mapsto (a_0, b_0)$:

$$a_0 = f_0 \cos\left(\frac{\pi}{4}\right) + f_1 \cos\left(\frac{3\pi}{4}\right) = \frac{f_0 - f_1}{\sqrt{2}},$$

$$b_0 = f_0 \sin\left(\frac{\pi}{4}\right) + f_1 \sin\left(\frac{3\pi}{4}\right) = \frac{f_0 + f_1}{\sqrt{2}}.$$

Theorem. The complexity of the fast Fourier transform algorithm for the vector of length $N = 2^n$ is $2N \log_2(N) = 2n2^n$.

Proof. We prove this Theorem by induction on n. For the basis of induction, $n = 1$, $N = 2$, we note that in this case the original formulas are used, with just 2 multiplications required, which is better than $2N \log_2(N)$.

Let us show the step of induction. Our induction assumption is that for the sequence of length 2^n, the complexity of FFT is $2n2^n$. Let us evaluate the complexity of FFT for a sequence of length 2^{n+1}. To compute Fourier coefficients a_p, b_p we first compute FFT for the subsequences **g** and **h** with complexity $2 \times 2n2^n$. Then for each of the 2^{n+1} Fourier coefficients, we need to carry out two multiplications. The total complexity is then

$$2 \times 2n2^n + 2 \times 2^{n+1} = 2(n+1)2^{n+1},$$

which is exactly what we need to prove.

One more remark is in order. When we express Fourier coefficients a_p and b_p through a_p^g, b_p^g, a_p^h, b_p^h, index p is in the range $0 \le p < 2^{n-1}$. However Fourier coefficients for **g** and **h** were only defined for the indices in the range $0 \le p < 2^{n-2}$. The following theorem explains how to evaluate the formulas for the Fourier coefficients for **g** and **h** with p in the range $2^{n-2} \le p < 2^{n-1}$.

Theorem. Let **g** be a vector of length 2^{n-1}. Let $p = 2^{n-1} - 1 - r$ with $0 \le r < 2^{n-2}$. Then

$$a_p^g = -a_r^g, \quad b_p^g = b_r^g.$$

Analogous relations hold for the Fourier coefficients of **h**.

Proof. Recall that Fourier coefficients a_p^g are given by the formula

$$a_p^g = \frac{1}{\sqrt{2^{n-2}}} \sum_{i=0}^{2^{n-1}-1} g_i \cos\left(\frac{(2p+1)(2i+1)\pi}{2^n}\right).$$

Substituting $p = 2^{n-1} - 1 - r$, we get

$$a_p^g = \frac{1}{\sqrt{2^{n-2}}} \sum_{i=0}^{2^{n-1}-1} g_i \cos\left(\frac{(2^n - 2r - 1)(2i + 1)\pi}{2^n}\right)$$

$$= \frac{1}{\sqrt{2^{n-2}}} \sum_{i=0}^{2^{n-1}-1} g_i \cos\left((2i + 1)\pi - \frac{(2r + 1)(2i + 1)\pi}{2^n}\right).$$

Using identities $\cos(2\pi + \alpha) = \cos(\alpha)$, $\cos(\pi + \alpha) = -\cos(\alpha)$, $\cos(-\alpha) = \cos(\alpha)$, we simplify the above to

$$a_p^g = -\frac{1}{\sqrt{2^{n-2}}} \sum_{i=0}^{2^{n-1}-1} g_i \cos\left(\frac{(2r + 1)(2i + 1)\pi}{2^n}\right) = -a_r^g.$$

The proof for b_p^g is left as an exercise.

In the case when $2^{n-2} \le p < 2^{n-1}$ we write p as $p = 2^{n-1} - 1 - r$ and we can calculate Fourier coefficients for $\mathbf{f}$ from the Fourier coefficients for $\mathbf{g}$ and $\mathbf{h}$ in the following way:

$$a_p = \frac{-a_r^g - a_r^h}{\sqrt{2}} \cos\left(\frac{(2p + 1)\pi}{2^{n+1}}\right) + \frac{b_r^g - b_r^h}{\sqrt{2}} \sin\left(\frac{(2p + 1)\pi}{2^{n+1}}\right),$$

$$b_p = \frac{b_r^g + b_r^h}{\sqrt{2}} \cos\left(\frac{(2p + 1)\pi}{2^{n+1}}\right) + \frac{a_r^g - a_r^h}{\sqrt{2}} \sin\left(\frac{(2p + 1)\pi}{2^{n+1}}\right).$$

Let us also express the trigonometric factors in the right hand side in terms of r:

$$\cos\left(\frac{(2p + 1)\pi}{2^{n+1}}\right) = \cos\left(\frac{(2^n - 2r - 1)\pi}{2^{n+1}}\right)$$

$$= \cos\left(\frac{\pi}{2} - \frac{(2r + 1)\pi}{2^{n+1}}\right) = \sin\left(\frac{(2r + 1)\pi}{2^{n+1}}\right).$$

In the last step we used the identity $\cos(\frac{\pi}{2} - \alpha) = \sin\alpha$.

Similarly,

$$\sin\left(\frac{(2p + 1)\pi}{2^{n+1}}\right) = \cos\left(\frac{(2r + 1)\pi}{2^{n+1}}\right).$$

Now for $2^{n-2} \leq p < 2^{n-1}$ with $p = 2^{n-1} - 1 - r$, we obtain:

$$a_p = \frac{-a_r^g - a_r^h}{\sqrt{2}} \sin\left(\frac{(2r+1)\pi}{2^{n+1}}\right) + \frac{b_r^g - b_r^h}{\sqrt{2}} \cos\left(\frac{(2r+1)\pi}{2^{n+1}}\right),$$

$$b_p = \frac{b_r^g + b_r^h}{\sqrt{2}} \sin\left(\frac{(2r+1)\pi}{2^{n+1}}\right) + \frac{a_r^g - a_r^h}{\sqrt{2}} \cos\left(\frac{(2r+1)\pi}{2^{n+1}}\right).$$

18 Quantum Fourier Transform

We have already introduced several standard transformations acting
on the spaces of qubits when we discussed quantum implementations
of classical computations. These quantum gates RAND, ROR, NOT
and CNOT are rather special transformations – all of these just per-
mute the basis vectors. To invoke full power of quantum computers
we shall need to go beyond the permutation transformations and introduce operations on 1- and 2-qubit spaces which are more general.

The first transformation we introduce is called the Hadamard's
transformation. It acts on a 1-qubit space in the following way:

$$H\left|0\right\rangle = \frac{1}{\sqrt{2}}\left|0\right\rangle + \frac{1}{\sqrt{2}}\left|1\right\rangle, \quad H\left|1\right\rangle = \frac{1}{\sqrt{2}}\left|0\right\rangle - \frac{1}{\sqrt{2}}\left|1\right\rangle.$$

The second standard family of quantum gates is rotation transformations R_α of a 1-qubit space:

$$R_\alpha\left|0\right\rangle = \cos\alpha\left|0\right\rangle - \sin\alpha\left|1\right\rangle, \quad R_\alpha\left|1\right\rangle = \sin\alpha\left|0\right\rangle + \cos\alpha\left|1\right\rangle.$$

The family of rotations, together with the Hadamard's transformation generate the orthogonal group $O(2)$ – every orthogonal transformation of a 1-qubit space is either a rotation, or a product of a
rotation with the Hadamard's transformation.

The last standard quantum transformation is the family of *controlled rotations* CR_α. These are transformations of a 2-qubit space
which act in the following way:

$$CR_\alpha\left|xy\right\rangle = \begin{cases} \left|xy\right\rangle & \text{if } x = 0, \\ \left|x\right\rangle R_\alpha\left|y\right\rangle & \text{if } x = 1. \end{cases}$$

Here we apply a rotation to the second quantum bit, but this rotation
is controlled by the value of the first quantum bit. The rotation is
applied only if the first bit has value 1, and is not applied if the first
bit has value 0.

In the standard basis of a 2-qubit space $\{\left|00\right\rangle, \left|01\right\rangle, \left|10\right\rangle, \left|11\right\rangle\}$, the

110

matrix of this transformation is:

$$
\mathrm{CR}_\alpha =
\begin{pmatrix}
1 & 0 & 0 & 0 \\
0 & 1 & 0 & 0 \\
0 & 0 & \cos\alpha & \sin\alpha \\
0 & 0 & -\sin\alpha & \cos\alpha
\end{pmatrix}
$$

Let us now define the *quantum Fourier transform* (QFT).

In the classical discrete Fourier transform the input is a vector $(f_0, f_1, \ldots, f_{N-1})$ and the output is the vector of Fourier coefficients $(a_0, a_1, \ldots, a_{N/2-1}, b_0, b_1, \ldots, b_{N/2-1})$. Quantum Fourier transform does the same thing, only the input and the output are n-qubits:

$$
\sum_{k=0}^{2^n-1} f_k \, |k_{n-1} \ldots k_1 k_0\rangle \mapsto
$$

$$
\sum_{p=0}^{2^{n-1}-1} a_p \, |p_0 p_1 \ldots p_{n-2}\rangle \, |0\rangle +
\sum_{p=0}^{2^{n-1}-1} b_p \, |p_0 p_1 \ldots p_{n-2}\rangle \, |1\rangle .
$$

Here $k_{n-1} \ldots k_1 k_0$ is the binary expression of integer k. In the right hand side it will be convenient for us to write the bits of p in the reverse order, with $p = p_{n-2} 2^{n-2} + \ldots + 2p_1 + p_0$ represented as $|p_0 p_1 \ldots p_{n-2}\rangle$. Since a_p coefficients are written in the Fourier transform sequence first, the value of the leading bit for them is 0, and the leading bit has value 1 for b_p coefficients.

The goal of this chapter is to present an implementation of quantum Fourier transform, using the standard 1-qubit and 2-qubit operations.

This implementation is built on the ideas of the fast Fourier transform. Just as it was the case with FFT, quantum Fourier transform is recursive. This means that we shall use QFT on $(n-1)$-qubits when constructing a QFT for n-qubits.

We shall first list 6 steps of the implementation of the quantum Fourier transform, and then go through these steps in detail. A nice feature of the QFT is that its implementation does not require any auxiliary memory.

Quantum Fourier Transform:

Step 1. Apply QFT to the first $(n-1)$ bits of an input n-qubit.
Step 2. Apply the Hadamard's transformation to the last bit.
Step 3. Apply CNOT to the last bit, controlled by the second last bit.
Step 4. Apply a sequence of rotations to the second last bit, controlled by the previous bits.
Step 5. Apply CNOT to the second last bit, controlled by the last bit.
Step 6. Apply CNOTs to the first $n-2$ bits, controlled by the second last bit.

We begin with the input n-qubit

$$\sum_{k=0}^{2^n-1} f_k \left| k_{n-1} \ldots k_1 k_0 \right\rangle .$$

First, following the ideas of fast Fourier transform, we splice the sequence $\mathbf{f}$ into two subsequences $\mathbf{g}$ and $\mathbf{h}$, with $\mathbf{g}$ corresponding to even values k and $\mathbf{h}$ corresponding to odd k. After this change in notations, our input qubit is written as

$$\sum_{s=0}^{2^{n-1}-1} g_s \left| s_{n-2} \ldots s_0 \right\rangle \left| 0 \right\rangle + h_s \left| s_{n-2} \ldots s_0 \right\rangle \left| 1 \right\rangle .$$

We perform Step 1 of our procedure, applying QFT to the left $n-1$ bits. The result of this operation is

$$\sum_{p=0}^{2^{n-2}-1} a_p^g \left| p_0 \ldots p_{n-3} \right\rangle \left| 00 \right\rangle + b_p^g \left| p_0 \ldots p_{n-3} \right\rangle \left| 10 \right\rangle$$

$$+ a_p^h \left| p_0 \ldots p_{n-3} \right\rangle \left| 01 \right\rangle + b_p^h \left| p_0 \ldots p_{n-3} \right\rangle \left| 11 \right\rangle .$$

Step 2 is to apply the Hadamard's transformation

$$H \left| 0 \right\rangle = \frac{1}{\sqrt{2}} \left| 0 \right\rangle + \frac{1}{\sqrt{2}} \left| 1 \right\rangle , \quad H \left| 1 \right\rangle = \frac{1}{\sqrt{2}} \left| 0 \right\rangle - \frac{1}{\sqrt{2}} \left| 1 \right\rangle$$

112

to the last bit, which produces

$$\sum_{p=0}^{2^{n-2}-1} \frac{a_p^g}{\sqrt{2}} |p_0 \dots p_{n-3}\rangle |00\rangle + \frac{a_p^g}{\sqrt{2}} |p_0 \dots p_{n-3}\rangle |01\rangle$$

$$+ \frac{b_p^g}{\sqrt{2}} |p_0 \dots p_{n-3}\rangle |10\rangle + \frac{b_p^g}{\sqrt{2}} |p_0 \dots p_{n-3}\rangle |11\rangle$$

$$+ \frac{a_p^h}{\sqrt{2}} |p_0 \dots p_{n-3}\rangle |00\rangle - \frac{a_p^h}{\sqrt{2}} |p_0 \dots p_{n-3}\rangle |01\rangle$$

$$+ \frac{b_p^h}{\sqrt{2}} |p_0 \dots p_{n-3}\rangle |10\rangle - \frac{b_p^h}{\sqrt{2}} |p_0 \dots p_{n-3}\rangle |11\rangle$$

$$= \sum_{p=0}^{2^{n-2}-1} \frac{a_p^g + a_p^h}{\sqrt{2}} |p_0 \dots p_{n-3}\rangle |00\rangle + \frac{a_p^g - a_p^h}{\sqrt{2}} |p_0 \dots p_{n-3}\rangle |01\rangle$$

$$+ \frac{b_p^g + b_p^h}{\sqrt{2}} |p_0 \dots p_{n-3}\rangle |10\rangle + \frac{b_p^g - b_p^h}{\sqrt{2}} |p_0 \dots p_{n-3}\rangle |11\rangle .$$

Step 3 is to perform CNOT on the last bit, controlled by the second last bit: $|x\, y\rangle \mapsto |x\ y \oplus x\rangle$, producing

$$\sum_{p=0}^{2^{n-2}-1} \frac{a_p^g + a_p^h}{\sqrt{2}} |p_0 \dots p_{n-3}\rangle |00\rangle + \frac{a_p^g - a_p^h}{\sqrt{2}} |p_0 \dots p_{n-3}\rangle |01\rangle$$

$$+ \frac{b_p^g + b_p^h}{\sqrt{2}} |p_0 \dots p_{n-3}\rangle |11\rangle + \frac{b_p^g - b_p^h}{\sqrt{2}} |p_0 \dots p_{n-3}\rangle |10\rangle .$$

The next Step 4 is the most intricate. We perform a sequence of rotations on the second last bit:

$$|0\rangle \mapsto \cos \alpha\, |0\rangle - \sin \alpha\, |1\rangle ,$$

$$|1\rangle \mapsto \sin \alpha\, |0\rangle + \cos \alpha\, |1\rangle .$$

The first rotation is in angle $\pi/2^{n+1}$. Then for each of the first $n-2$ bits $p_0 \dots p_{n-3}$ we perform a rotation of the second last bit in angle $\pi/2^{n-j}$, which is controlled by bit p_j. This rotation is performed only if $p_j = 1$, and we do nothing if $p_j = 0$. Alternatively, we can interpret this controlled rotation as a rotation of the second last bit in the angle $p_j \pi / 2^{n-j}$.

Combining these rotations together, we will get a rotation in angle

$$\alpha = (1 + 2p_0 + 2^2 p_1 + \ldots + 2^{n-2} p_{n-3}) \frac{\pi}{2^{n+1}} = \frac{(2p+1)\pi}{2^{n+1}}.$$

Below we are going to abbreviate $|p_0 \ldots p_{n-3}\rangle$ to $|p\rangle$. The outcome of Step 4 is

$$\sum_{p=0}^{2^{n-2}-1} \frac{a_p^g + a_p^h}{\sqrt{2}} \cos\left(\frac{(2p+1)\pi}{2^{n+1}}\right) |p\rangle |00\rangle - \frac{a_p^g + a_p^h}{\sqrt{2}} \sin\left(\frac{(2p+1)\pi}{2^{n+1}}\right) |p\rangle |10\rangle$$

$$+ \frac{a_p^g - a_p^h}{\sqrt{2}} \cos\left(\frac{(2p+1)\pi}{2^{n+1}}\right) |p\rangle |01\rangle - \frac{a_p^g - a_p^h}{\sqrt{2}} \sin\left(\frac{(2p+1)\pi}{2^{n+1}}\right) |p\rangle |11\rangle$$

$$+ \frac{b_p^g + b_p^h}{\sqrt{2}} \sin\left(\frac{(2p+1)\pi}{2^{n+1}}\right) |p\rangle |01\rangle + \frac{b_p^g + b_p^h}{\sqrt{2}} \cos\left(\frac{(2p+1)\pi}{2^{n+1}}\right) |p\rangle |11\rangle$$

$$+ \frac{b_p^g - b_p^h}{\sqrt{2}} \sin\left(\frac{(2p+1)\pi}{2^{n+1}}\right) |p\rangle |00\rangle + \frac{b_p^g - b_p^h}{\sqrt{2}} \cos\left(\frac{(2p+1)\pi}{2^{n+1}}\right) |p\rangle |10\rangle$$

$$= \sum_{p=0}^{2^{n-2}-1} \left(\frac{a_p^g + a_p^h}{\sqrt{2}} \cos\left(\frac{(2p+1)\pi}{2^{n+1}}\right) + \frac{b_p^g - b_p^h}{\sqrt{2}} \sin\left(\frac{(2p+1)\pi}{2^{n+1}}\right) \right) |p\rangle |00\rangle$$

$$+ \left(\frac{a_p^g - a_p^h}{\sqrt{2}} \cos\left(\frac{(2p+1)\pi}{2^{n+1}}\right) + \frac{b_p^g + b_p^h}{\sqrt{2}} \sin\left(\frac{(2p+1)\pi}{2^{n+1}}\right) \right) |p\rangle |01\rangle$$

$$+ \left(\frac{b_p^g - b_p^h}{\sqrt{2}} \cos\left(\frac{(2p+1)\pi}{2^{n+1}}\right) - \frac{a_p^g + a_p^h}{\sqrt{2}} \sin\left(\frac{(2p+1)\pi}{2^{n+1}}\right) \right) |p\rangle |10\rangle$$

$$+ \left(\frac{b_p^g + b_p^h}{\sqrt{2}} \cos\left(\frac{(2p+1)\pi}{2^{n+1}}\right) - \frac{a_p^g - a_p^h}{\sqrt{2}} \sin\left(\frac{(2p+1)\pi}{2^{n+1}}\right) \right) |p\rangle |11\rangle.$$

On Step 5 we apply CNOT to the second last bit, controlled by the last bit: $|x\ y\rangle \mapsto |x \oplus y\ y\rangle$, yielding

$$\sum_{p=0}^{2^{n-2}-1} \left(\frac{a_p^g + a_p^h}{\sqrt{2}} \cos\left(\frac{(2p+1)\pi}{2^{n+1}} \right) + \frac{b_p^g - b_p^h}{\sqrt{2}} \sin\left(\frac{(2p+1)\pi}{2^{n+1}} \right) \right) |p\rangle\,|00\rangle$$

$$+ \left(\frac{a_p^g - a_p^h}{\sqrt{2}} \cos\left(\frac{(2p+1)\pi}{2^{n+1}} \right) + \frac{b_p^g + b_p^h}{\sqrt{2}} \sin\left(\frac{(2p+1)\pi}{2^{n+1}} \right) \right) |p\rangle\,|11\rangle$$

$$+ \left(\frac{b_p^g - b_p^h}{\sqrt{2}} \cos\left(\frac{(2p+1)\pi}{2^{n+1}} \right) - \frac{a_p^g + a_p^h}{\sqrt{2}} \sin\left(\frac{(2p+1)\pi}{2^{n+1}} \right) \right) |p\rangle\,|10\rangle$$

$$+ \left(\frac{b_p^g + b_p^h}{\sqrt{2}} \cos\left(\frac{(2p+1)\pi}{2^{n+1}} \right) - \frac{a_p^g - a_p^h}{\sqrt{2}} \sin\left(\frac{(2p+1)\pi}{2^{n+1}} \right) \right) |p\rangle\,|01\rangle\,.$$

Our final step is to apply controlled negations CNOT to each of first $n-2$ bits, all controlled by the second last bit. When writing the final outcome, we will split it into two parts. Whenever the second last bit is zero, we will keep p as a summation variable, and when value of the second last bit is 1, we will replace p with a new variable r. The bits of r are all going to be negated when we apply CNOT, and we will denote the result of negation by $|\bar{r}\rangle = |\bar{r}_0\bar{r}_1\ldots\bar{r}_{n-3}\rangle$. We then get the following expression:

$$\sum_{p=0}^{2^{n-2}-1} \left(\frac{a_p^g + a_p^h}{\sqrt{2}} \cos\left(\frac{(2p+1)\pi}{2^{n+1}} \right) + \frac{b_p^g - b_p^h}{\sqrt{2}} \sin\left(\frac{(2p+1)\pi}{2^{n+1}} \right) \right) |p\rangle\,|00\rangle$$

$$+ \left(\frac{b_p^g + b_p^h}{\sqrt{2}} \cos\left(\frac{(2p+1)\pi}{2^{n+1}} \right) - \frac{a_p^g - a_p^h}{\sqrt{2}} \sin\left(\frac{(2p+1)\pi}{2^{n+1}} \right) \right) |p\rangle\,|01\rangle$$

$$+ \sum_{r=0}^{2^{n-2}-1} \left(\frac{b_r^g - b_r^h}{\sqrt{2}} \cos\left(\frac{(2r+1)\pi}{2^{n+1}} \right) - \frac{a_r^g + a_r^h}{\sqrt{2}} \sin\left(\frac{(2r+1)\pi}{2^{n+1}} \right) \right) |\bar{r}\rangle\,|10\rangle$$

$$+ \left(\frac{a_r^g - a_r^h}{\sqrt{2}} \cos\left(\frac{(2r+1)\pi}{2^{n+1}} \right) + \frac{b_r^g + b_r^h}{\sqrt{2}} \sin\left(\frac{(2r+1)\pi}{2^{n+1}} \right) \right) |\bar{r}\rangle\,|11\rangle\,.$$

Recalling the formulas of the fast Fourier transform, we recognize the first two coefficients as a_p and b_p with $0 \leq p < 2^{n-2}$, while the last two coefficients are a_p and b_p with $2^{n-2} \leq p < 2^{n-1}$, $p = 2^{n-1} - 1 - r$. The above expression simplifies to

$$\sum_{p=0}^{2^{n-2}-1} \left(a_p \left| p \right\rangle \left| 00 \right\rangle + b_p \left| p \right\rangle \left| 01 \right\rangle \right) + \sum_{\substack{p=2^{n-2} \\ p=2^{n-1}-1-r}}^{2^{n-1}-1} \left(a_p \left| \bar{r} \right\rangle \left| 10 \right\rangle + b_p \left| \bar{r} \right\rangle \left| 11 \right\rangle \right).$$

When p with $2^{n-2} \leq p < 2^{n-1}$, $p = 2^{n-1} - 1 - r$, is written as an $n - 1$ bit integer in binary form, the leading bit of p is equal to 1, and the remaining $n - 2$ bits are negations of the bits of r, $\left| p \right\rangle = \left| \bar{r} \right\rangle \left| 1 \right\rangle$, so the two parts of the summation may be combined, taking p to be now an $n - 1$ bit integer, $\left| p \right\rangle = \left| p_0 p_1 \ldots p_{n-3} p_{n-2} \right\rangle$:

$$\sum_{p=0}^{2^{n-1}-1} a_p \left| p_0 p_1 \ldots p_{n-2} \right\rangle \left| 0 \right\rangle + b_p \left| p_0 p_1 \ldots p_{n-2} \right\rangle \left| 1 \right\rangle.$$

We see that as a result of this sequence of steps, we obtained precisely the quantum Fourier transform of the input n-qubit.

Exercise: Show that the complexity of the quantum Fourier transform for n-qubits (the number of the quantum gates used) is n^2.

We see that QFT has a quadratic complexity (as a function of the number of bits), versus exponential complexity for FFT.

19 Shor's Algorithm

It is now time to combine all the threads we developed and describe Shor's quantum algorithm for factorization of large integers.

Let p, q be two large secret prime numbers, and $N = pq$ be their product. Our goal is to determine the values of p and q, given the value of N. Solving this problem will break RSA cryptosystem. As we have seen, we can determine p and q if, in addition to N, we know the order of the group $\mathbb{Z}_N^*$ of invertible remainders mod N. We shall denote the number of elements in $\mathbb{Z}_N^*$ by $M = (p-1)(q-1)$.

The idea is to determine M by finding orders of elements in $\mathbb{Z}_N^*$. By Lagrange's Theorem, the order of each element is a divisor of M, hence, computing the orders of elements in this group will give us divisors of M.

Let us illustrate this idea with the following example. Let us select primes $p = 36013$ and $q = 51199$, with $N = pq = 1843829587$. Numbers $p - 1$ and $q - 1$ factor as follows: $p - 1 = 2^2 \times 3 \times 3001$, $q - 1 = 2 \times 3 \times 7 \times 23 \times 53$. In practice, these factorizations will not be known as the primes p, q themselves are not given. Still, for illustration purposes it is helpful to have these factorizations as they shed light on the orders of elements in $\mathbb{Z}_N^*$. The order of this multiplicative group is $M = (p-1)(q-1) = 1843742376 = 2^3 \times 3^2 \times 7 \times 23 \times 53 \times 3001$.

Suppose we have a way to compute orders of elements in this group. Let us list orders of a few elements:

g	order of g	factorization of the order
2	13360452	$2^2 \times 3 \times 7 \times 53 \times 3001$
3	21949314	$2 \times 3 \times 23 \times 53 \times 3001$
5	13360452	$2^2 \times 3 \times 7 \times 53 \times 3001$
7	5797932	$2^2 \times 3 \times 7 \times 23 \times 3001$
11	43898628	$2^2 \times 3 \times 23 \times 53 \times 3001$
13	153645198	$2 \times 3 \times 7 \times 23 \times 53 \times 3001$

Here we do not list the order of $g = 4$ since being the square of 2, its order is a half of the order of 2. In an abelian group, the order of a product is a divisor of the least common multiple of the orders of factors. For this reason, we only list prime elements g.

Again, in a real situation, factorization of orders will not be available, but it is instructive to look at these. Even if we do not have factorizations of orders available, we can still effectively compute the least common multiple of the orders of elements since

$$\text{LCM}(a,b) = \frac{ab}{\text{GCD}(a,b)}$$

and the greatest common divisor may be computed using the Euclidean algorithm.

It follows from Lagrange's Theorem that M is divisible by the least common multiple of the orders of elements. For the elements we listed above, the least common multiple is

$$R = 307290396 = 2^2 \times 3 \times 7 \times 23 \times 53 \times 3001 = \text{LCM}(p-1, q-1).$$

In a setting of cryptography, numbers $p-1$ and $q-1$ will have large prime factors, otherwise there are known non-quantum methods for breaking RSA. This implies that $\text{LCM}(p-1, q-1)$ will differ from M by a fairly small factor. Since $N/M \approx 1$, we can determine this factor from $N/R = \frac{1843829587}{307290396} \approx 6.00028$, hence $M = 6R$.

This tells us that if we can determine orders of a few elements in the group $\mathbb{Z}_N^*$, we will be able to factor N. For the rest of this chapter we will focus on the problem of finding the order m of a given element g in $\mathbb{Z}_N^*$.

We begin by setting the number of qubits that will be required for the quantum algorithm. Choose n such that $2^{n-1} < N < 2^n$. Quantum algorithm that we are going to describe will be operating with $3n$-qubits.

We begin with a $3n$-qubit initialized to $|00\ldots0\rangle$.

Step 1. We apply Hadamard's transformation H to each of the first $2n$ qubits:

$$H\,|0\rangle \ldots H\,|0\rangle\,|0\ldots0\rangle = \frac{1}{\sqrt{2}}\big(\,|0\rangle + |1\rangle\,\big) \ldots \frac{1}{\sqrt{2}}\big(\,|0\rangle + |1\rangle\,\big)\,|0\ldots0\rangle$$

$$= \frac{1}{2^n} \sum_{k=0}^{2^{2n}-1} |k\rangle\,|0\ldots0\rangle.$$

Step 2. For the given remainder g in $\mathbb{Z}_N^*$ consider a function

$$f : B_{2n} \longrightarrow B_n,$$

$$f(k) = g^k \mod N, \quad 0 \le k < 2^{2n}.$$

Let us point out that function f is periodic with period equal to the order of element g. Since this order (which we want to determine) does not exceed 2^n, we are computing this function over a large number of its periods. Classically this computation is not feasible, but can be done with a quantum computer due to massive parallelism of quantum computations.

The classical computation f has a quantum implementation T_f, which is a linear transformation of the space of $3n$-qubits, where the first $2n$ bits are interpreted as input bits, and the last n bits as output. Since $N < 2^n$, we have a sufficient number of bits to record any remainder mod N.

Apply T_f to the qubit constructed in Step 1:

$$T_f \left(\frac{1}{2^n} \sum_{k=0}^{2^{2n}-1} |k\rangle \, |0 \ldots 0\rangle \right) = \frac{1}{2^n} \sum_{k=0}^{2^{2n}-1} |k\rangle \, | \, g^k \mod N \, \rangle.$$

Step 3. Perform the measurement of the last n bits. The result of this measurement is probabilistic. We will observe one of the values h, which is a power of $g \mod N$. Our $3n$-qubit will collapse to a state, where the last n bits will assume value $h \mod N$. What will happen with the first $2n$ bits? All the values of k with $g^k \ne h \mod N$ will disappear. Still, the resulting state will involve a sum, since there is more than one value of k for which $g^k = h \mod N$. Denote by s the smallest such value of k. As m is the order of g, we have

$$h = g^s = g^{s+m} = g^{s+2m} = g^{s+3m} = \ldots = g^{s+(L-1)m} \mod N,$$

where L is the smallest integer greater or equal to $(2^{2n} - s)/m$. Then the quantum state obtained as a result of this measurement can be written as

$$\frac{1}{\sqrt{L}} \sum_{j=0}^{L-1} |s + jm\rangle \, |h\rangle.$$

We see that the coefficients in this state form a sequence, which is periodic with period m (which is to be determined):

$$f_k = \frac{1}{\sqrt{L}} \begin{cases} 1, & \text{if } k = s \bmod m, \\ 0, & \text{otherwise.} \end{cases}$$

Step 4. Apply quantum Fourier transform to the first $2n$ bits. This will result in the state

$$\sum_{r=0}^{2^{2n-1}-1} a_r |r\rangle |0\rangle |h\rangle + b_r |r\rangle |1\rangle |h\rangle,$$

where the values of a_r and b_r have spikes at integer multiples of the frequency $\omega = 2^{2n}/m$ (see the exercise in the chapter on discrete Fourier transform).

Step 5. Perform the measurement of the values of the first $2n - 1$ bits. The outcome of the measurement is probabilistic, with the probability of observing value r being $a_r^2 + b_r^2$. With a high probability, the observed value of r will correspond to a spike, and hence r will be close to an integer multiple of $\omega = 2^{2n}/m$.

We record the observed value of r and repeat Steps 1–5 several times, collecting observations $r_1, r_2, \ldots, r_\ell$ (here ℓ is a small number). This completes the quantum part of the algorithm.

Let us describe how we can get the value of m from the observed values $r_1, r_2, \ldots, r_\ell$. As we pointed out, $r_1, \ldots, r_\ell$ are close to being integer multiples of $\omega = 2^{2n}/m$. Let us assume that $r_i \approx k_i \omega$. We are hoping that $\text{GCD}(k_1, \ldots, k_\ell) = 1$. This is quite realistic to expect. Indeed, the probability that two random large integers are relatively prime is $6/\pi^2 = 0.6079\ldots$. For ℓ random large integers this probability is given by the inverse of the so-called zeta-function: $1/\zeta(\ell)$. For example, for $\ell = 6$, the probability that 6 random integers do not share a common factor is $1/\zeta(6) = 945/\pi^6 = 0.9829\ldots$.

Our goal is to determine the value of ω, as we can easily find m from it: $m = 2^{2n}/\omega$. Although $r_1, \ldots, r_\ell$ are essentially integer multiples of ω, we cannot use the standard Euclidean algorithm to find the value of ω, after all ω itself is not an integer. Instead, we shall describe a version of the Euclidean algorithm which finds an *approximate GCD*, and can even work for real numbers.

We are going to explain this method through an example. Let us take $N = 989$ as the number to be factored. We want to find the order m of $g = 2$ in the multiplicative group $\mathbb{Z}^*_{989}$ using Shor's algorithm. Since $989 < 2^{10} = 1024$, we set $n = 10$. After Step 2 of Shor's algorithm, we get the qubit:

$$\frac{1}{2^{10}} \sum_{k=0}^{2^{20}-1} |k\rangle \left| 2^k \mod 989 \right\rangle.$$

On Step 3, we measure the value of the last 10 bits, with a random value being observed. Let us assume that the result of this measurement is $h = 2^{50} = 41 \mod 989$. Then only the terms where $2^k = 41 \mod 989$ will survive in the resulting sum. These values of k form an arithmetic progression $k = 50,\ 50 + m,\ 50 + 2m, \dots$. Hence the sequence of coefficients in the qubit is now periodic with period m, and applying quantum Fourier transform on Step 5, we will get the state

$$\sum_{r=0}^{2^{19}-1} a_r |r\rangle |0\rangle |h\rangle + b_r |r\rangle |1\rangle |h\rangle,$$

with the Fourier coefficients having spikes at values of r which are close to an integer multiple of $\omega = 2^{20}/m$. To illustrate this, we plot

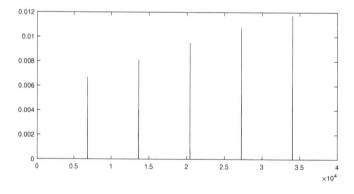

the graph of $a_r^2 + b_r^2$ for r from 0 to 40,000. We should understand that the values of the coefficients of the qubit stored on a quantum computer cannot be directly accessible, and we can plot this graph only because we use a small value of N, and can compute the Fourier coefficients with a classical computer. This graph is a plot of probability of observing each value of r if we perform the measurement of the first 19 qubits. We clearly see the spikes on this graph, which means that some values of r are much more likely to be measured than others. Locations of these spikes do not depend on the observed value h.

Here we also list probabilities of observing values of r near two of the peaks:

r	$a_r^2 + b_r^2$	r	$a_r^2 + b_r^2$
251926	0.0000071	435767	0.000054
251927	0.0000124	435768	0.000091
251928	0.0000271	435769	0.000186
251929	0.0000991	435770	0.000567
251930	0.0125868	435771	0.008652
251931	0.0001465	435772	0.002382
251932	0.0000329	435773	0.000373
251933	0.0000141	435774	0.000145
251934	0.0000078	435775	0.000076

We can see from these tables that some values of r are much more likely to be observed. Suppose we run the quantum algorithm twice and get $r_1 = 435771$ as observed value on the first run, and $r_2 = 251930$ on the second run. We expect these values to be approximate multiples of $\omega = 2^{20}/m$.

Let us run the Euclidean algorithm for finding the greatest common factor of r_1 and r_2. We divide r_1 by r_2 with a remainder:

$$r_3 = r_1 - r_2 = 183841.$$

The remainder r_3 is always lower than the divisor r_2, however we do not expect it to be lower by many orders of magnitude. Indeed, for $0 < c < 1$ the probability that $r_3 < cr_2$ is equal to c.

Let us continue with the Euclidean algorithm.

$$r_4 = r_2 - r_3 \quad = 68089,$$
$$r_5 = r_3 - 2 \times r_4 = 47663,$$
$$r_6 = r_4 - r_5 \quad = 20426,$$
$$r_7 = r_5 - 2 \times r_6 = 6811,$$
$$r_8 = r_6 - 3 \times r_7 = 7.$$

At the last step we notice that the value of the remainder $r_8 = 7$ is several orders of magnitude lower than the order of the divisor $r_7 = 6811$. We interpret this as $r_8 \approx 0$, which is the termination condition for the search of the approximate GCD. Hence r_7 is the approximate GCD of r_1 and r_2. Let us determine the corresponding integer multiples:

$$\frac{r_1}{r_7} = \frac{435771}{6811} = 63.98\ldots, \qquad \frac{r_2}{r_7} = \frac{251930}{6811} = 36.98\ldots$$

Then we get a more precise value of the base frequency w:

$$w \approx \frac{435771}{64} \approx 6808.921\ldots, \qquad w \approx \frac{251930}{37} \approx 6808.918\ldots.$$

Now we can determine the value of the order m of element $g = 2$ in $\mathbb{Z}_{989}^*$ from $w = 2^{20}/m$:

$$m \approx \frac{2^{20}}{w} \approx \frac{1048576}{6808.92} \approx 154.0003\ldots$$

We conclude that the order of $g = 2$ is $m = 154$.

By Lagrange's Theorem, the size $M = (p-1)(q-1)$ of the group $\mathbb{Z}_{989}^*$ is an integer multiple of the order of any element, $M = mK$. Since the magnitude of $M = (p-1)(q-1)$ is comparable to $N = pq = 989$, we can approximate the unknown factor K as

$$K = \frac{M}{m} < \frac{N}{m} = \frac{989}{154} \approx 6.42\ldots$$

Taking $K = 6$, we obtain the value of $M = 6m = 924$.

Since $N = pq$ and $M = (p-1)(q-1) = pq - p - q + 1$, we get that

$$p + q = N - M + 1 = 989 - 924 + 1 = 66.$$

Finally, we find the prime factors p, q as the roots of the quadratic equation

$$X^2 - (p+q)X + pq = 0,$$

$$X^2 - 66X + 989 = 0,$$

$$p = \frac{66 + \sqrt{66^2 - 4 \times 989}}{2} = 43, \quad q = \frac{66 - \sqrt{66^2 - 4 \times 989}}{2} = 23,$$

and we can verify obtained factorization:

$$43 \times 23 = 989.$$

Exercise. Show that in our description of Shor's algorithm, Step 3 may be skipped.

20 Appendix: What is not in this Book?

Complex numbers.

The proper way to describe quantum states is with complex, and not real numbers. Let us give a brief introduction to complex numbers here. The starting point to complex numbers is introduction of the imaginary number i, which is the square root of -1, meaning that we postulate $i^2 = -1$. Then the general form for a complex number is $a + bi$, where a and b are real numbers, called the real part and the imaginary part of the complex number. Let us give examples how we can add and multiply complex numbers:

$$(2 + 5i) + (1 - 3i) = (2 + 1) + (5 - 3)i = 3 + 2i,$$

$$(2 + 5i) \times (1 - 3i) = 2 \times 1 + 2 \times (-3i) + 1 \times 5i + 5 \times (-3) \times i^2 =$$
$$2 - 6i + 5i - 15 \times (-1) = 17 - i.$$

A new operation with complex numbers is *conjugation*, which is switching the sign of the imaginary part. It is denoted with a bar:

$$\overline{2 + 5i} = 2 - 5i.$$

Conjugation has the following properties:

$$\overline{z + w} = \overline{z} + \overline{w}, \qquad \overline{z \times w} = \overline{z} \times \overline{w},$$

for any two complex numbers z, w.

There is also an analogue of the absolute value, which is called the *norm* of a complex number:

$$|a + bi| = \sqrt{a^2 + b^2}.$$

The norm of a non-zero complex number is a positive real number. The following relation, which is easy to verify, gives a connection between conjugation and norm:

$$z \times \overline{z} = |z|^2.$$

This relation allows us do define division of complex numbers:

$$\frac{1}{z} = \frac{\overline{z}}{|z|^2}.$$

Let us give an example:

$$\frac{2+5i}{1+3i} = \frac{(2+5i)\overline{(1+3i)}}{|1+3i|^2} = \frac{(2+5i)(1-3i)}{|1+3i|^2} = \frac{17-i}{1^2+3^2} = \frac{17}{10} - \frac{1}{10}i.$$

A *Hermitian* vector space consists of vectors, whose components are complex numbers. The dot product in a Hermitian space is defined as follows:

$$\begin{pmatrix} z_1 \\ z_2 \\ \dots \\ z_n \end{pmatrix} \cdot \begin{pmatrix} w_1 \\ w_2 \\ \dots \\ w_n \end{pmatrix} = \overline{z}_1 w_1 + \overline{z}_2 w_2 + \dots + \overline{z}_n w_n.$$

With this definition, the dot product of a non-zero complex vector with itself is a positive real number:

$$\mathbf{u} \cdot \mathbf{u} = |z_1|^2 + |z_2|^2 + \dots + |z_n|^2,$$

for

$$\mathbf{u} = \begin{pmatrix} z_1 \\ z_2 \\ \dots \\ z_n \end{pmatrix}.$$

This allows us to define the length of a complex vector as $|\mathbf{u}| = \sqrt{\mathbf{u} \cdot \mathbf{u}}$. A qubit is a complex vector:

$$(a+bi)\,|0\rangle + (c+di)\,|1\rangle.$$

If we take into consideration rotational polarization states of a photon, together with the linear polarization then quantum states of a photon are modelled by complex 1-qubits.

More generally, an n-qubit is expressed as a complex vector

$$\sum_{k=0}^{2^n-1} z_k \,|k\rangle$$

with the condition that the length of such a vector is equal to 1:

$$\sum_{k=0}^{2^n-1} |z_k|^2 = 1.$$

When we perform a measurement on this n-qubit, the probability of observing value k is $|z_k|^2$, which is a non-negative real number.

When we discuss evolution of quantum states in complex setting, real orthogonal matrices are replaced with matrices of complex numbers, satisfying the condition $A^{-1} = \overline{A}^T$. Here conjugation of a matrix is carried out by taking a conjugate of each entry. A complex matrix satisfying this condition is called *unitary*.

Many functions can be extended to complex numbers. A particularly important example is the complex exponential function. Exponentials of complex numbers are defined with the following *Euler's formula:*

$$e^{a+bi} = e^a \left(\cos(b) + i \sin(b) \right).$$

A complex version of discrete Fourier transform is based on complex exponentials. It turns out that complex version of DFT is easier to work with, compared to the real case. Fourier coefficients of a complex sequence $(f_0, f_1, \ldots, f_{N-1})$ are computed as follows:

$$c_p = \frac{1}{\sqrt{N}} \sum_{k=0}^{N-1} f_k e^{-2\pi i k p / N}.$$

The inverse Fourier transform reconstructs the original sequence in a very similar fashion:

$$f_k = \frac{1}{\sqrt{N}} \sum_{p=0}^{N-1} c_p e^{2\pi i k p / N}.$$

We can use these formulas to build the quantum Fourier transform in complex setting with unitary quantum gates.

Physical realizations of qubits.

In fact, quantum computers may be based not only on photonics, but may be built with several other physical systems. These include:

- Spins of electrons,
- Nitrogen atoms embedded into a diamond crystal,
- Trapped ions,
- Quantum dots,
- Josephson junctions (superconducting circuits).

Other quantum algorithms.

Quantum algorithms are notoriously difficult to design. To-date, only a small number of them is known, and of those, nothing is as spectacular as Shor's algorithm.

Let us list here a few examples of quantum algorithms:

- Grover's algorithm for solving equations $f(x) = a$,
- Finding repeated values (solving equations $f(x) = f(y)$),
- Computing Jones' polynomial invariants of knots.

Let us outline the idea for the Grover's algorithm. Suppose $f(x)$ is a classical computation, $f : B_n \to B_k$. For a given a in B_k we want to solve equation $f(x) = a$. Let us assume, for simplicity, that this equation has a unique (unknown) solution $x = r$ in B_n. We set $N = 2^n$.

We begin with the initial n-qubit

$$\mathbf{v} = \frac{1}{\sqrt{N}} \sum_{k=0}^{N-1} |k\rangle .$$

To solve the problem, we need to amplify the coefficient in the qubit that corresponds to the term that we seek: $\mathbf{u} = |r\rangle$.

Let us find the angle β between vectors $\mathbf{v}$ and $\mathbf{u}$. Using dot product we determine that $\cos\beta = \frac{1}{\sqrt{N}}$. This value is close to zero, thus β is close to $\pi/2$. Writing $\beta = \pi/2 - \alpha$, and using the fact that $\cos(\pi/2 - \alpha) = \sin\alpha$, we determine that $\alpha = \arcsin(1/\sqrt{N}) \approx 1/\sqrt{N}$.

Grover was able to construct a rotation transformation $R_{2\alpha}$ in angle 2α of a plane spanned by vectors $\mathbf{u}$ and $\mathbf{v}$. It turns out that we do not need to know $\mathbf{u} = |r\rangle$ to construct $R_{2\alpha}$, it is enough to know that $f(r) = a$. Every time we apply this rotation, the angle between the current quantum state and $\mathbf{u}$ gets reduced by 2α. If k is an integer satisfying $2k + 1 \approx \pi\sqrt{N}/2$, then after k rotations, we get a vector, which is close to $\mathbf{u} = |r\rangle$. Performing the measurement, we will obtain r with a high probability.

Grover's algorithm requires a multiple of $\sqrt{N}$ steps. This is an improvement over the classical situation, where on average $N/2$ steps are required to find the solution of equation $f(x) = a$ by a direct search. If f does not have any nice properties which could give us a better method for solving this equation, it is clear that classically this problem cannot be solved in $\sqrt{N}$ steps. Thus in case of the Grover's algorithm, we can prove that quantum methods are more effective than classical.

Still, the speed-up to $\sqrt{N}$ from N is not as impressive as in the case of Shor's algorithm, where the speed-up is logarithmic.

Alternative quantum computational schemes.

In addition to quantum circuits considered in this book, there exist quantum computational schemes based on different principles.

In *adiabatic* quantum computers, quantum states are implemented with Josephson junctions, a collection of superconducting circuits cooled to a fraction of 1 Kelvin above absolute zero. These circuits are controlled by a magnetic field.

For each magnetic field configuration, there exist many possible quantum states of superconducting circuits, corresponding to different energy levels of the system. We are interested in finding the *ground state*, i.e. the state with the lowest energy. If we use an analogy with a vibrating string, the ground vibrational state corresponds to the base frequency mode. Let us emphasize that the "particle" we consider here is not the string, but rather its vibration. Thus a non-vibrating string represents absence of the particle, and the particle in its lowest energy state is the base mode. Here we study the smile separately from the Cheshire Cat.

One has to design a magnetic field configuration for which the ground state gives a solution to a problem we wish to solve. The difficulty is that if we just create this configuration of the magnetic field, there is no guarantee, that our quantum processor will be in the ground state. The strategy is then the following: we first create a very simple magnetic field configuration, for which the ground state is known (and corresponds to the zero value of the n-qubit). We initialize the quantum processor to the zero ground state. Then we begin to gradually change the magnetic field configuration, morphing it into the final configuration which should produce the solution to our problem. While we are changing the magnetic field, the quantum state of the processor remains in the ground energy state, which evolves in response to the changing magnetic field. The final ground state yields the solution of the problem.

Yet another model for quantum computations is given by *topological* quantum computers. In this model quantum computations are done by braiding the trajectories of *anyons*. Anyons are not actual particles, but rather excitations in a system of electrons. Such a system changes its state when two anyons circle each other. There is hope that topological quantum computers will be more stable since the result of the computation depends only on the configuration of anyon trajectories and is tolerant to small perturbations.